I·

Oven Drying

II

Oven Drying

The Best Way to Preserve Foods

IRENE CROWE

SHEED ANDREWS AND McMEEL, INC.

SUBSIDIARY OF UNIVERSAL PRESS SYNDICATE

KANSAS CITY

IV

First printing, June 1976
Second printing, September 1976

Library of Congress Cataloging in Publication Data
Crowe, Irene, 1921-
    Oven drying, the best way to preserve foods.

    Includes index.
        1. Food—Drying.   2. Cookery (Dried foods)
I. Title.
TX609.C68   641.4'4   76-14427
ISBN 0-8362-0664-9

This book is

lovingly dedicated

to family and friends

This book is

to family and friends.

C
O
N
T
E
N
T
S

Contents

VIII

Preface

So far, information on drying foods has been scant and in some instances difficult to follow because of complicated instructions and time-consuming details. The information contained in this book has been carefully selected and tested for its simplicity of preparation and carried through so that the reconstituted product meets eye and taste approval.

In addition to providing instructions on the preparation of dried fruits, vegetables, meats, fish, herbs, and nuts, I have added a chapter on miscellaneous foods. This section incorporates delicious recipes for cookies and candies that are tasty and nutritious, granola (cereal), egg drying, pectin (necessary to jell fruits), and vinegar to preserve foods.

Foods will keep for long periods of time, if carefully prepared and as carefully packaged, using minimal equipment and products at hand.

Writing this book has given me great satisfaction and I hope in turn you will use it to your enjoyment and good health.

Caution to Readers

The instructions in *Oven Drying: The Best Way to Preserve Foods* are safe and effective if followed with ordinary care. However, a few cautionary words at the start should be kept in mind as you read this book.

In some electric ovens, the upper broiler element may stay partially on, even with a bake setting. Do *not* attempt to remove the upper element since there is danger of electrocution. Instead, aluminum foil with the shiny side up on a rack on the uppermost shelf position should be used to deflect the direct heat from the broiler element. If you have a gas oven without an automatic gas shutoff valve, you should check the oven from time to time to see if the flame is still on. Regardless of the type of oven you have, do not leave the oven unattended when it is in use. If part of the drying must be completed overnight or while you are away, cover the product and freeze until you are ready to reprocess.

All foods with mold on them should be discarded. Simple removal of the moldy part is not sufficient removal of all the unseen toxic micro-organisms produced from molds.

Sodium bisulfite should be used only in the amounts indicated. It should not be drunk nor its fumes inhaled, and it should be kept out of the reach of children. Do not confuse sodium bisul*fite* with sodium bisul*fate* (the latter is toxic in solution and is an irritant to eyes and skin).

Oven pasteurization is recommended for all fruits and small-cut vegetables in accordance with the procedure described in the *General Information* section.

Chpt. 1.

General Information

Sun drying is the oldest method of food preservation. Our ancestors, through trial and error, learned to use the sun's rays to help them, and they depended on sun-dried food to support, nourish, and sustain them from one growing season to another.

Today, the drying processes have changed and man can easily dry food with a minimum of work or fuss uncomplicated by the inconstancy of nature. His need, though different, is still as basic as that of his forefathers.

Rising costs and food shortages during the out-of-season period have made food drying a sensible practice. More and more people are drying their surplus food, including leftovers that can be easily and conveniently added to other foods later. Escalating prices and shortages have also given us a chance to reevaluate our needs and life styles. We are now attempting to work with nature to avoid contaminating the soil and atmosphere by returning to earlier styles of production and preservation of foods. The economy and self-satisfaction derived from this project are well worth checking into. It is fun!

1

There are many ways to dry foods. They range from the simple sun drying to oven drying and the more complex dehydrators. All deal with the same principles: the evaporation of liquids, attractive appearance, taste, and retention of nutritive values. Success depends on the ripeness and quality of the product and on quick handling and processing. Continuing care should be exercised throughout. Carelessness in details may result in an inferior product that might prove costly or a total loss.

In this book we will deal with the simple method of oven drying as a way of preserving, with excellent results, excess foods that one might raise or buy. These instructions are applicable and can be used interchangeably with other dryers and dehydrators. Just follow your instruction manual carefully.

A food should be of excellent quality and in prime condition for its particular season. Use only firm ripe fruits. Refrigerate the excess until the oven is available for drying. Always start with a product at its peak in color, taste, and vitamin potency. Wilted or old fruits and vegetables will result in a poor product. Blemished or bruised fruits are susceptible to spoilage and will not keep well. If only parts of fruits or vegetables are salvageable, the affected areas may be removed and the balance diced for use in stews, soups, or fruit compotes. Carefully label these packages and hold under refrigeration or in a freezer.

There are many ways to prepare foods for drying. Some foods need only to be washed and sliced, diced, or cut into pieces of desired size before placing in the oven to dry. Other foods need pretreatment to keep them in their prime condition. Darkening of foods occurs when certain chemicals within the food unite

2

with oxygen. To prevent this, fruits and vegetables are placed into a solution strong enough to protect foods from changing color during the drying process.

There are many solutions used for preserving foods. Each will give its maximum effect for the particular food used. The most commonly used methods are salt, sugar-salt, and vinegar-salt solutions. In addition, sodium bisulfite, a GRAS (Generally Recognized as Safe) substance can be used on many foods to prevent them from discoloring. However, it should not be used on meat, enriched and whole-grain cereals, legumes, or any other foods rich in thiamin as it will destroy that vitamin content. Bisulfiting takes the place of sulfuring and leaves no taste. Treated foods may be kept under refrigeration without discoloration. Sodium bisulfite may be added to the water before steaming or blanching or can be incorporated in the steaming process. It can be purchased at your local wine-making supply store or at photography supply houses. Check under *Pretreatment of Foods* for your choice of solutions.

Simple blanching is recommended for some foods. Prepared foods are placed into simmering water for a specified time. Steaming is another excellent way of retaining food values and good color. Whatever method you use, follow the instructions carefully.

Oven drying is simple and effective. An oven temperature which is too low may cause foods to sour, while a temperature which is too high will harden the outer casing and prevent inner moisture from escaping. The temperatures for oven drying different foods will vary. But to prevent spoilage, the heat must be sustained at at least 140° F. for more than half the total drying time of any spoilable food. (Some greens and herbs, which burn easily, may be dried at lower temperatures.)

Do not overload the trays, as this prevents air circulation and lengthens drying time. Food should be arranged in a single layer and exposed on the top and bottom. Rotate trays up and down from time to time so that food will dry evenly. Also rotate them from front to back. Fruits which are very juicy, such as apples, apricots, berries, and watermelon, may drip excessively from tray to tray and have the effect of washing off the pretreatment solution on foods placed on lower trays. This effect may be overcome by placing paper toweling beneath the trays during the first half hour or so when the dripping is most pronounced. If paper toweling is used, however, it must be watched carefully to be sure it does not burn. Aluminum foil may also be placed on the bottom of the oven to catch spills throughout the drying process.

Throughout the oven drying procedure for all foods, the oven door must remain partially open. The oven door on a gas range should be propped open 8 inches, on an electric range half an inch. In some electric ovens, the upper broiler element may stay partially on, even with a bake setting. Do *not* attempt to remove the upper element, since there is danger of electrocution. Instead, aluminum foil with the shiny side up on a rack on the uppermost shelf position should be used to deflect the direct heat from the broiler element.

If you have a gas oven without an automatic gas shutoff valve, you should check the oven from time to time to see that the flame is still on. Regardless of the type of oven you have, do not leave the oven unattended when it is in use. If part of the drying must be done overnight or while you are away, cover the product and freeze until you are ready to reprocess. In any case, toward the end of the drying time it is wise to lower the oven temperature. A little experimentation will help determine how this can be accomplished in your particular oven. If your oven retains temperature for long periods, you may be able to turn off the oven and finish the drying in that way. Or you may find it more effective for your oven to maintain a low temperature but leave the oven door completely open toward the end.

Drying is complete when fruit feels dry and leathery on the outside and slightly moist on the inside. Vegetables are dry when they are brittle, tough, and rattle on the trays. Foods are softer when they are still warm. For this reason foods should be cooled before they are checked for dryness. If in doubt, leave the food on the tray a little longer, reducing the temperature, but do not overdry. Food that overheats near the end of the drying time will scorch easily. A dried food is reduced in shape and size and is firm and brittle. Reconstituting will return it to its natural state and volume.

Although the terms drying and dehydrating are commonly used interchangeably, according to the USDA commercial dehydration involves a highly sophisticated process that cannot be duplicated at home and reduces the moisture content to only 2.5 to 4 percent water. Dried foods, on the other hand, still contain roughly 10 to 20 percent water, depending on whether they are vegetables or fruits. We will therefore refer to this process as oven drying rather than dehydration. The moisture content of dried food suggests yet another way of testing for dryness. A kitchen scale can be a useful way to check for dryness. If, for example, a cup of the food being processed weighed

10 ounces before drying, it should weigh between 1 and 2 ounces when it is dry.

Oven pasteurization is recommended for all fruits and small-cut vegetables after the oven drying process is completed in order to remove all spoilage elements that may not have been killed off in the oven drying process. Small-cut vegetables should be heated for an additional 10 minutes and fruits for an additional 15 minutes in an oven preheated to 175° F.

Store the food in airtight plastic or glass containers. Sealable plastic bags, now available commercially, are ideal containers, as they take up little space in the cupboard and can be transported easily on camping trips. Check foods periodically for any sign of moistness, mold, or infestation. Return to the oven for further drying if moisture is present. If mold or infestation is present, discard the food. Properly prepared and stored, food should keep up to a year.

Some foods do not require reconstitution. Dried fruit is delicious eaten as it is or chopped and mixed in cereals, cookies, etc. Reconstituted, it is excellent in desserts, purees, or stews. Presoaking or overnight soaking is necessary for some foods. The amount of water depends on the amount of food and the degree of dryness. Follow individual recipes for instructions on reconstitution. Because most vegetables contain starch, which is not easily digested if eaten dry, they should be soaked in water before they are eaten. Follow the instructions given under each particular food.

Chpt. **2.**

gres fm fg . 7 to 9 = 3 fp

Equipment

Because standard kitchen equipment is used, the expenses are minimal. The greatest outlay is the cost of the steamer, thermometer, and materials for making drying racks. A kitchen scale may also be useful.

If you do not have a steamer you may improvise by placing a metal colander (plastic will melt) over 2 cups of simmering water. Cover and steam the food according to the time specified for each recipe. Foods may also be parboiled before drying. This method is not as good as steaming because the nutrients escape into the water. Steaming is by far the easiest and best way to blanch foods, with less loss of food values, resulting in better dried foods. Double boilers are not advisable as the air does not circulate freely.

A high-quality oven thermometer is a handy kitchen gadget that will keep you informed of the oven temperature at a glance; oven dials are frequently inaccurate guides to the true temperature of an oven. Be sure you are familiar with the temperature ranges of your particular oven. Check and recheck. If you lower the temperature in your gas oven, check periodically to be sure that the flame does not go out. You can also spot check the oven heat by holding some of the drying food momentarily or by touching the oven grids. If food or grids are too hot to handle, the temperature is over 150° F. If foods or grids are too cool, the temperature is under 150° F. Oven regulation is easy: If oven temperatures are too high, lower the temperature or lower the oven door. The amount of heat will go down appreciably. A little patience and vigilance soon results in an easy adjustment to fluctuating temperatures.

Shallow wooden trays should be constructed so that air will circulate freely through the oven. They should be at least 1½ to 2 inches shorter than the oven dimensions on all four sides to provide for proper ventilation. They may be made to fit on the grids of your oven racks or as a single unit to fit inside the oven. There should be at least 3 inches of space above and below the racks. Aromatic or green wood should not be used for these trays.

For the bottoms, trays may be fitted with wooden slats placed ¼ to ½ inch apart. Do not use galvanized screen, because it has been treated with zinc and cadmium. These metals will react with the acids in food and could cause food poisoning. Also avoid fiberglass mesh, as tiny splinters of fiberglass may be freed easily and impregnate the food.

Other materials which can be used for bottoms are ungalvanized screen, stainless steel perforated metal, and nylon, polyester, or polypropylene-mesh products which have a high melting point. If wooden trays are used they may be brushed with salad oil or a vegetable oil spray-on may be used to protect the wood and to make tray cleaning easier.

To prevent foods from sticking to the trays, two

layers of cheesecloth or cotton netting should be tacked on top of the trays. These may be washed and reused.

After every use, wash and dry all equipment and store in a well-ventilated area.

from 10
gres to pg. 11 ... =2/58.

chpt 3.

Pretreatment of Foods

Food must be excellent in quality and at its finest in maturity. The time elapsed between picking, cleaning, and blanching should be minimal. While many foods may be dried, pretreatment is essential to hold color and flavor, to aid in softening the tissues for faster drying, and to stop enzyme action to prevent further ripening. There are several methods of treating foods for drying and these methods are a matter of personal choice.

Sodium Bisulfite provides by far the best treatment for preservation of foods which do not have a high thiamin content. It is not advisable for meats, cereals or legumes. Foods will retain their color, drying time is lessened, and reconstituted food returns to its former shape in less time. Prepared food may be treated and kept under refrigeration without discoloration. These holding qualities allow a higher finishing temperature and will reduce drying time by almost a third. Sodium bisulfite is tasteless and it also eliminates the tedious process of sulfuring. It may be purchased at your local wine-making supply store or photography supply houses. Follow package instructions. _WARN-_

ING: Avoid use of bisulfites with copper and iron. Use glass, plastic, aluminum, stainless steel, or porcelain bowls. Sodium bisulfite should be used only in the amounts indicated. Since it should not be drunk nor its fumes inhaled, it should be kept out of the reach of children. Do not confuse sodium bisulf*ite* with sodium bisulf*ate* (the latter is a toxic solution and is an irritant to eyes and skin).

Cold Sulfite Solution: Use 3 tablespoons of sodium bisulfite to 1 gallon of water. Soak fruits and vegetables 1 minute. Drain and spread on trays.

Sodium Bisulfite Steaming: This eliminates one step if you blanch. Add 1 to 2 teaspoons sodium bisulfite to 2 cups of water. Steam required time. This step will retard the loss of carotene and vitamins and will retain the color and flavor of the product.

Salt Solution: Use 4 to 6 tablespoons salt to a gallon of water. Submerge fruit 10 minutes.

Salt-Sugar Solution: Use 6 tablespoons salt and ½ cup of sugar to each gallon of water. Submerge fruit for 10 minutes.

Salt-Vinegar Solution: Use 2 tablespoons salt and 2 tablespoons vinegar to ½ gallon of water. Submerge food for 10 minutes.

Remember, if large amounts of food are processed, change solutions occasionally. The efficacy of the solution might not be up to par and the food will remain untreated. Also, regardless of the treatment used, if the food is not at its peak in quality it will not result in a good dried product.

gres for pg 12 to 13 ... = 2 pgs.   chpt 4.

Blanching

Some foods are merely washed, cut into desired portions, and placed in the oven to dry. Other foods must be blanched before drying. Blanching sets the color, hastens the drying time, and prevents undesirable changes in flavor during storage. Blanched vegetables when reconstituted require less soaking before they are cooked and also have a better flavor. Steam blanching is preferred to water blanching, as steam retains nutrients that are lost in the water.

For water blanching, foods are simply immersed into boiling water for the required time, stopping the enzyme action responsible for color changes, odors, and textures. It also destroys insect infestation. Small amounts of food are slowly added to keep the simmering water from chilling. Food is immersed for the required time and cooled before it is placed on the trays for drying. Leafy vegetables may also be processed this way. Drain off any excess moisture and pat dry. Use paper toweling if necessary. The disadvantages to this method are color bleeding and loss of vitamins and minerals.

Steaming is by far the best method, as vitamins,

minerals, sugars, and other food values are more fully preserved. Steaming inactivates or destroys the natural enzymes responsible for changes and loss of vitamins and will also deter bacterial formation. With steaming, drying occurs more quickly, reconstitution is swift, and the food will keep longer.

Rapid handling will help in maintaining an excellent product. Steam the food until tender but firm so that vitamins and other substances are retained. Otherwise the product will have an off flavor or odor. You may want to add 1 to 2 teaspoons of sodium bisulfite to the water. (See *Pretreatment of Foods.*) It helps immeasurably in the steaming process.

Steam blanching is essential for certain fruits. Fruits such as firm berries like blueberries, cherries, grapes, or plums have relatively tough skins with a waxlike coating. To remove this waxy substance and to permit inside moisture to come to the surface and evaporate, the skins are cracked or "checked" by blanching.

Do not use a double boiler to steam food. If you do not have a steamer, place a wire strainer or metal colander (plastic will melt) over a pan of simmering water. Cover the food during the blanching process. Remember that the product must be translucent when done and some foods will attain this condition more quickly than others.

Do not overblanch or your product will be tough, fibrous, and difficult to handle. Food will stick to trays, drying time will be extended, and reconstituted food may lack both shape and substance. Loss of flavor or odor could result in an insipid product.

5.

goes from pg 14 to 52 = 39 pp.

Drying Fruits
(39 fruits listed)

Fruits should be dried when they are at their peak in quality and flavor. Many are harvested in this state while others are picked and gradually ripened to this prime condition. Fruits are classified under the categories of stone bearing, pomes, berries, citrus, capsules, and pepos. They contain sugar, starch, pectin, cellulose, and other organic acids. During the drying process the starches are converted to sugar, and for this reason sugar should be used with caution in cooking reconstituted fruit. Oversweetening will overpower the fruit flavor and taste.

Many fruits are picked from trees while others are harvested from the ground. Fruits such as apples, cherries, peaches, nectarines, and apricots should be picked when ripe. Many show this ripeness by a softness at the stem end. Pears and persimmons are two of the fruits that can be picked while immature, allowed to ripen gradually, and dried at their peak flavor. Fruits such as prunes and figs may be harvested immediately after they have dropped to the ground, having reached their full potential. These fruits should be picked daily to avoid spoilage and contamination through decay or insect infestation. When ripe, they should be held under refrigeration and then dried as quickly as possible.

Soft fruits and culls should be reserved for cooling drinks, jams, jellies, preserves, sauces, and fruit leathers. These may be incorporated into breakfast treats or luncheon and dinner desserts. Many of these are delightful when mixed and add color and variety to the diet. They are also economical when in season. Turn the extras into future servings and savings. During the winter months these delicious treats will brighten and warm many a cold and wintery day.

Fruits contain vitamins A and C as well as other protective minerals. Grapes, bananas, dates, figs, and prunes are most nutritious as they contain a large amount of sugar. Watermelon, oranges, lemons, grapefruits, and grapes contain a large amount of moisture which helps in cleansing the kidneys in addition to their nutritive benefits. Some fruits act as a stimulant and add variance to a bland diet. Others are served just for their subtle and tantalizing flavors.

Fruits should be carefully washed and checked for blemishes and decay. Peaches can be defuzzed or peeled. To defuzz, rub a cloth over the peel until most of the fuzz is gone. Many prefer peeling, as the skin sometimes sets the teeth on edge. Some types of pears are grainy and are better peeled. The peelings may be reserved and simmered until tender. Put through 2 strainings to catch grainy particles. Use as a puree in your favorite recipe. Whole fruit may be dipped into boiling water to check the skins or steamed to this condition. This process is necessary to permit inner moisture to evaporate.

Most fruits are treated by one of several methods to

prevent oxidation. Fruits may be blanched or immersed in a solution to reserve color and to prevent darkening. Check the section *Pretreatment of Foods.* ~p10 + 11.
The use of bisulfite is by far the most superior method as it will maintain color and will hasten drying time and reconstitution later.

Syrup-blanching of fruit before drying gives a sweetened product. Simmer fruits for 10 minutes in a mixture consisting of 1 cup sugar to 3 cups water. After simmering, allow the fruit to stand in this solution for 10 minutes, then drain. Care should be maintained not to overblanch as the product will lack firmness, will adhere to the trays, and will take longer to dry.

Pat dry to absorb any surface moisture and place on trays. Paper toweling may be placed under the trays and aluminum foil may be placed on the bottom of the oven to catch drips. Remove paper toweling when fruit firms up. Watch paper toweling carefully to see that it doesn't burn. Fruit should be spread in a single layer to avoid overcrowding. If there is not enough air circulation the fruit will sour or drying time will be prolonged. Heat should never rise above 165° F. If this cannot be controlled, lower the oven door. Use a thermometer to regulate heat cycle. The oven door on a gas range must be propped open 8 inches, on an electric range half an inch.

Fruit that is drying properly should feel cooler and more moist than the oven air. If it is the same temperature it is drying too fast. Trays should be rotated. When dry, fruit should be leather-like on the outside and slightly moist inwardly. Allow to cool when checking for dryness. Do not add fresh fruits to partially dried fruits as fruit will take longer to dry.

Dried fruit should be carefully packaged in airtight,

moistureproof bags or containers and checked occasionally. If mold or infestation occurs, discard the product.

Dried fruit may be eaten in its dried state or can be reconstituted by barely covering with water and refrigerating for several hours or overnight. Fruit should resume its former appearance. Fruit may be cut up into salads, baked in cookies and cakes, or simmered into a sauce. Raisins, dates, and figs will blend better if they are soaked in boiling water for 5 minutes.

Various fruits may be combined into fruit leathers. Crushed nuts may be added to the recipe or for a variation use oatmeal as a base for the pureed fruit, top with crushed nuts, and dry to a crisp cracker. These make delightful eating and are nutritious as well. Children will enjoy these treats, which are superior to any commercial product.

① Apples

Almost any apple may be dried, although the early apples should not be used. They lack the flavor and firmness that result in a good dried product and are best used in applesauce and apple butter. Dried apples are delicious and may be used for snacking or added to salads when reconstituted.

Mature apples have a developed fragrance and flavor and are one of the most healthful foods on the market. Immature apples and fruits contain a large amount of starch and should be stewed to make them more digestible. Fruit keeps at its peak when the temperature is cool and the air is somewhat moist.

Apples discolor easily and should be treated before

drying. Discoloration does not affect the product, however, and discolored apples may still be used. To prevent this discoloration, steam blanch prepared fruit for 3 to 5 minutes or until there is no moisture in the center. Or use any method of pretreatment you desire. See the section *Pretreatment of Foods.*

Place peeled, cored, ringed, sliced, or diced fruit into desired solution for the required time and then onto trays. Place the trays into a 140° F. oven for 2 hours. The oven door on a gas range must be propped open 8 inches, on an electric range half an inch. Increase heat to 160° F. for 2 hours and return to 140° F. until the fruit is dry, pliable, and springy. Turn apples to dry evenly.

Aluminum foil may be placed at the oven bottom and paper toweling under trays to catch drips. Watch paper toweling carefully to see that it doesn't burn. Remove toweling when fruit firms up. Store in airtight containers when dry.

Drying time: 5-7 hours for apple rings.

To reconstitute: Add 1 cup dehydrated apples to 2 cups water. Presoak for 1 hour, then bring to a boil. Simmer covered for about 15 minutes or until tender. Add seasonings and sugar. Use for pies, sauces, or stewed fruits.

Apricots

When apricots are ripe enough to eat, but not soft, they are just right for drying. The fruit bruises easily and for convenient handling it is best to cut the fruit in half. (It may be quartered to hasten drying time.) Remove the pits and drop fruit into prepared solution.

See the section *Pretreatment of Foods*. If you choose water blanching, place the whole fruit in boiling water for 4-5 minutes then cut and pit. Do not overblanch, as soft apricots are difficult to handle and the fruit will stick to the trays. Arrange the fruit on trays, cut side up, and place into 150° F. oven for 3 hours. The oven door on a gas range must be propped open 8 inches, on an electric range half an inch. Turn fruit, reduce heat to 140° F., and continue to dry for 6-12 hours or longer until the fruit is dry. If the fruit begins to scorch, lower the oven temperature to 100° F. or leave the oven door completely open. You may place paper toweling under trays and aluminum foil at bottom of oven to catch drips. Watch paper toweling carefully to see that it doesn't burn. Remove paper toweling when food firms up. Dried apricots should be soft but pliable and leathery, with no moistness in the center when cut. Store in airtight containers in a cool dry place when dry.

Drying time: 9-15 hours for halves.

To reconstitute: Place 1 cup of dried fruit and 2 cups of water in a covered bowl and refrigerate for several hours or overnight.

Stewed fruit: Add water ¼ inch above fruit. Simmer uncovered for 15 minutes or until fruit is plump and tender but not mushy. Add sugar to taste. Lemon juice may also be added.

Avocados

The avocado, sometimes called alligator pear, usually reaches the market in an unripe state. It's a slightly pear-shaped fruit and when ripe peels easily. The outside colors range from green to black. It is sometimes marked with a brown scabbing. Fruit which has a bright appearance and is just beginning to soften is best. It has a nutlike buttery taste that adapts itself to many foods. Avocados are excellent in salads and dips and are delicious served alone with just a touch of dressing or a sprinkle of lime or lemon juice. The avocado bruises and discolors easily and should be handled carefully as bruising does affect the quality of the fruit.

Peel, pit, and cut into desired slices or dices. Drop into desired solution. See the section *Pretreatment of Foods*. Place in a 140° F. oven with the door on a gas range propped open 8 inches, on an electric range half an inch. Store in airtight containers when dry.

To reconstitute: Add water to cover and refrigerate for several hours before using.

Bananas

Ripe fruit should have a fresh appearance, attractive color, and firm pulp. If the fruit is tipped with green it is only partially ripe and may be a little tart. Let stand at room temperature until yellow, flecked with brown. The starch has then been converted into natural sugar.

Cut into ¼ inch slices and place into desired solution. See the section *Pretreatment of Foods*. Drain. Place on trays in a 140° F. oven until chewy and leathery. The oven door on a gas range must be propped open 8 inches, on an electric range half an inch. If the fruit sticks to the trays do not attempt to loosen it until it is firm. Insert a knife carefully under slices and loosen gently. Store in airtight containers

when dry. This makes a delicious and chewy snack!
Drying time: 5-6 hours for slices.
To reconstitute: Add 1 cup bananas to 2 cups
water. Soak at least 2 hours before using.

Berries

The term berry is usually applied to any small juicy
fruit that is generally edible. In botany it is applied to
pulpy fruits with thin skins that are fleshy throughout,
contain one or more seeds, and have no stone or pit.
Under this classification, cranberries, currants, grapes,
the hip of the rose, and tomatoes could all be listed as
true berries. All berries can be dried.

Berries should have a fresh, clean appearance and be
free from dirt, blemishes, and moisture. They should
be plump and fragrant with a deep color throughout.
These deliciously flavored fruits perish rapidly and
must be treated quickly but gently to insure keeping
qualities. Fruit should not be allowed to stand at
room temperature or overnight. Berries lose their flavor rapidly and should be bought in small quantities
that can be handled or processed immediately. Otherwise cover and refrigerate. Larger amounts may be
spread out in layers and covered tightly to retain their
fine qualities.

Do use fresh berries frequently during their season.
Experiment with various fruits by combining them for
breakfast treats, salads, or desserts. Try different herbs
and spices for a tantalizing gourmet treat. Fresh pineapples and strawberries touched with a bit of mint is a
very delightful dish. Use these fruits with sherbets,
ices, bombes, whips, and liqueurs. Dry some for later

21

use.

Berries should never be cleaned under running water as they bruise easily. Place them in a bowl of water and carefully lift them with cupped fingers, allowing the soil and sand to drift to the bottom. Two light washings are sufficient. Do not pour water off berries as the sand will remain in the bowl and adhere to the fruit.

Strawberries, raspberries, and other soft fruits are simply sorted, washed, and dried. Larger fruit may be cut into smaller pieces to facilitate quicker drying.

Fruit with thin skins should be steam blanched for 2 minutes or longer to crack the skins and then placed on trays.

Fruit may be placed in a solution for better keeping qualities. This step will enhance color, speed up the drying process, and berries will reconstitute more quickly. See the section *Pretreatment of Foods.*

Place fruit in a single layer on trays into a 120° F. oven for 2 hours. Then raise the heat to 140° F. Redistribute fruit on trays and separate any clusters. You may place paper toweling under trays and aluminum foil at oven bottom to catch drips. Watch carefully that paper doesn't burn. Remove paper toweling when fruit firms up. Berries make good leathers. See the section on *Leathers.* Store in airtight containers when dry.

Drying time: 2-5 hours.

To reconstitute: Cover with water. Place the berries in the refrigerator for several hours or overnight. Reconstituted berries are not as tasty as fresh, canned, or frozen ones. To powder for use as a flavoring or in syrup, cut dried fruit into small pieces and continue drying until very brittle. Place in a blender or grind into a powder.

BLACKBERRIES. —

Ripe blackberries should be firm, well developed, and have a lustrous, glossy black color. They should be free from caps, mold, decay, dirt, disease, and insects. Green or red berries are not yet ripe and should be avoided.

Place on trays in a 120° F. oven for 2 hours. Then raise the temperature to 140° F., redistributing fruit on trays and separating any clusters. Store in airtight containers when dry.

To reconstitute: Cover with water. Place the berries in the refrigerator for several hours or overnight.

Young blackberry leaves that have been sorted and washed can be dried and used as a tea. Place on the lower rack of an open oven at low heat until dry. Store in airtight containers in a cool dry place.

Pour about 6 cups of boiling water over a pinch of the leaves for a very pleasant-tasting drink. Steep for about 10 minutes. Do not add milk, although sugar may be added if desired.

BLUEBERRIES. —

Blueberries have a light dewy color. They are sweet, with a hint of tartness. They have tender skins, with seeds that are small and almost inconspicuous. Huckleberries, sometimes confused with blueberries, are darker with larger seeds. They are less desirable for this reason. Ripeness in blueberries is indicated by the color which may range from blue, to blue-black, to purple. Good quality blueberries should be fresh, clean, and plump with a deep color throughout. They must be free from moisture. Overripe fruit is dull, lifeless, soft, watery, and sometimes shriveled.

Place the blueberries on trays in a 120° F. oven for

2 hours. Then raise the temperature to 140° F., redistributing fruit on trays and separating any clusters. Store in airtight containers when dry.

To reconstitute: Cover with water. Place the berries in refrigerator for several hours or overnight.

BOYSENBERRIES. —

These are black berry-like fruits that are similar to raspberries in appearance and flavor but are larger, with bigger seeds. In recipes they may be used in place of blackberries or raspberries.

Place boysenberries on trays in a 120° F. oven for 2 hours. Then raise the oven temperature to 140° F., redistributing fruit on trays and separating any clusters. Store in airtight containers when dry.

To reconstitute: Cover with water. Place the berries in the refrigerator for several hours or overnight.

DEWBERRIES. —

Dewberry is a popular name for certain brambles and blackberries. Dewberries closely resemble the trailing blackberry in leaves, stems, and fruit. They may be substituted in any blackberry recipe.

Place dewberries on trays in a 120° F. oven for 2 hours. Then raise the oven temperature to 140° F., redistributing berries on trays and separating any clusters. Store in airtight containers when dry.

To reconstitute: Cover with water. Place the berries in the refrigerator for several hours or overnight.

ELDERBERRIES. —

The elderberry is commonly used in wine. Collect and sort through for fully ripe berries. They are delicious when dried.

Place the elderberries on trays in a 120° F. oven for

2 hours. Then raise the oven temperature to 140° F., redistributing berries and separating any clusters. Store in airtight containers when dry.

To reconstitute: Cover with water. Place in refrigerator for several hours or overnight.

The flower clusters may be dipped in batter and fried like a fritter.

LOGANBERRIES. —
These are a cross between blackberries and red raspberries. They have the shape of a blackberry and the flavor of a tart red raspberry. Loganberries are used in beverages, as a breakfast fruit, and in desserts. They do not keep well and should be stored in a cool, dry place. Wash thoroughly before using.

Place the loganberries on trays in a 120° F. oven for 2 hours. Then raise the oven temperature to 140° F., redistributing berries on trays and separating any clusters. Store in airtight containers when dry.

To reconstitute: Cover with water. Place in refrigerator several hours or overnight.

RASPBERRIES. —
A perennial plant of the genus *Rubus*, raspberries are cultivated for their delicious fruits. They are closely related to blackberries and dewberries. Red raspberries have a more delicate flavor than black raspberries. Fruit should be fresh, plump, and well developed with a natural aroma.

Capped berries indicate early harvesting. Fruit that has an off color or green cells will not be as flavorful as the mature berry. Avoid stained boxes and soft fruits as they are indicative of decay. Fruit should be chilled and washed just before using. Avoid prolonged soaking or handling.

Place the raspberries on trays in 120° F. oven for 2 hours. Then raise the oven temperature to 140° F., redistributing berries on trays and separating any clusters. Store in airtight containers when dry.

To reconstitute: Cover with water. Place in refrigerator for several hours or overnight.

STRAWBERRIES. —

The strawberry is one of the most popular fruits. Its welcome appearance in the spring heralds the coming fruit season. The berries' tart but pleasant tang livens up the palate. Their color brightens up meals and they blend beautifully with other fruits. They are said to be a cleansing agent. Tea may be made from the young tender leaves.

Check berries to see if they have been roughly handled, were picked green, are overripe, or if there is any mold in the center. These berries are a poor buy. Strawberries should be kept chilled and washed just before they are used as they are tender and will perish and deteriorate rapidly.

Place the strawberries on trays in a 120° F. oven for 2 hours. Then raise the oven temperature to 140° F., redistributing fruit on trays and separating clusters. Store in airtight containers when dry.

To reconstitute: Cover with water. Place in refrigerator for several hours or overnight.

Cantaloupes

Cantaloupes must be allowed to ripen on the vine to reach their peak flavor. Maturity and quality is indicated when stems separate under thumb pressure or when cracks appear all around the stem. This usually denotes full sweet flavor and fruit that has a definite cantaloupe aroma. The coarse tan or grayish color of the netting and veining should be well developed.

Avoid fruit which is hard, slick, or poorly netted as it is usually tough and flavorless. Overripe fruit is characterized by a pronounced yellowing of the rind with flesh that is soft, watery, and tasteless.

Remove rind. Cut fruit into ½ inch slices or smaller pieces and place into desired solution. See the section *Pretreatment of Foods.*

Drain, then place on trays in an oven set at 140° F. The oven door on a gas range must be propped open 8 inches, on an electric range half an inch. You may place paper toweling under trays and aluminum foil at the oven bottom to catch drips. Watch carefully that paper doesn't burn. Remove paper toweling when fruit firms up. Drying time is longer as cantaloupes have a high water content. Finish drying at a low temperature or leave the oven door completely open. Store in airtight containers when dry.

Drying time: 2 or more hours for ½ inch pieces.

To reconstitute: Place in water to cover. Cantaloupe chips make a better snack food and are tastier than the reconstituted fruit.

Cherries

Mature, top-quality cherries have a bright, fresh, plump appearance and a good color. They are juicy and have an excellent flavor. Both sweet and sour cherries may be dried. Wash and pick cherries. Discard any cherries that are soft, overripe, or shriveled. Wormy cherries will float to the top of the water when soaked.

Cherries must be blanched to speed up the drying process. Follow the instructions for thin-skinned fruit given in the section on *Berries*. Place on trays in an oven set at 120° F. and gradually raise the temperature to 150° F. The oven door on a gas range must be propped open 8 inches, on an electric range half an inch. You may place paper toweling under trays and aluminum foil at the bottom of the oven to catch drips. Watch carefully that paper doesn't burn. Remove paper toweling when fruit firms up. Dry until fruit is leathery but not tacky. Store in airtight containers.

Drying time: 2 or more hours.

To reconstitute: Add 1 cup fruit to 1 cup water. Use less water for pies.

Citrus

Citrus is a genus of plants of the rue family, often thorny, with fruits that have a spongy or firm peel, and pulpy, juicy flesh. These include the grapefruit, lemon, lime, orange, and tangerine. They contain many vitamins but are especially valuable for their high vitamin C content.

Fruit should be washed and placed in boiling water for 5 minutes to loosen the skin. Cool, then remove skin in quarters, exercising care not to cut into pulp. The white membrane can then be easily removed. Segment fruit.

Citrus may be processed two ways. One way is to

place segments into sodium bisulfite solution or another desired solution. Follow the instructions given in the section *Pretreatment of Foods*. Drain. Pierce segments with a fork before placing on trays. Evaporation will be speedier. Remove seeds if there are any.

Dry at a low temperature of 140° F. The oven door on a gas range must be propped open 8 inches, on an electric range half an inch. You may place paper toweling under trays and aluminum foil at oven bottom to catch drips. Remove paper toweling when fruit firms up. Store in airtight containers when dry.

Or 3 cups of segmented fruit may be placed into a simmering syrup solution consisting of 4 cups sugar to 4 cups water. Simmer fruit for 3 minutes. Let cool in syrup 10 minutes, then drain and place on trays. Pierce fruit with a fork and remove seeds if there are any. Dry at a low temperature of 140° F. The oven door on a gas range must be propped open 8 inches, on an electric range half an inch. Store in airtight containers when dry.

Drying time: 3 or more hours.

This makes an excellent snack food. Reconstituted fruit is bitter.

CITRUS PEEL. —

Cut or slice the peels into desired pieces and steam for 10 minutes to remove bitter taste. If you use sodium bisulfite, the peel will retain its bright color. Drain and place on trays and dry at a low temperature of 140° F. Keep the oven door completely open. Store in airtight containers when dry.

Drying time: 2 or more hours.

Citrus peel is excellent when served with tea. Use to season or spice food. It can be ground if desired.

Or cut into desired slices or pieces and combine with 4 cups water to which ¼ teaspoon of baking soda has been added. Simmer covered for 20 minutes stirring occasionally. Rinse. Drain and place on trays in a low oven to dry. Store in airtight containers when dry. The peel will be less fragrant but less biting with this method.

GRAPEFRUIT. —

Grapefruit is the largest citrus fruit, with a bright yellow or pink skin and an acid pulp that is slightly bitter. The color of the fruit is either white or a pinkish-red with a fleshy pulp. The latter is much sweeter than the white. Grapefruits are good for breakfast as a fruit or drink, and delicious when added to salads. Good quality grapefruits are mature, firm, thin-skinned, smooth textured, and springy to the touch.

To dry: Grapefruit may be processed two ways. One way is to place segments into sodium bisulfite solution or another desired solution. Follow the instructions given in the section *Pretreatment of Foods* and the heading *Citrus*. Drain. Pierce segments with a fork before placing on trays. Evaporation will be speedier. Remove seeds if there are any. Dry at a low temperature of 140° F. Store in airtight containers when dry.

Or 3 cups of segmented fruit may be placed into a simmering syrup solution consisting of 4 cups sugar to 4 cups water. Simmer fruits for 3 minutes. Let cool in syrup 10 minutes, then drain and place on trays. Pierce fruit with a fork and remove seeds if there are any. Dry at a low temperature of 140° F. Store in airtight containers when dry.

KUMQUATS. —

A small elongated citrus fruit, the kumquat is related

to the orange. This small fruit's outer skin is faintly sweet and biting and its inner flesh can run from mildly sweet to the tartness of a lemon. Kumquats are aromatic and have a pleasant refreshing taste. They can be served whole, sliced fresh in salads, or candied. Or they may be made into marmalades, jellies, and preserves. Choose fruit which is firm and heavy for its size with skins that are bright and unblemished. Wash before using.

To dry: Do not skin fruit. Cut in half. Kumquats may be processed two ways. One way is to place segments into sodium bisulfite solution or another desired solution. Follow the instructions given under the section *Pretreatment of Foods* and the heading *Citrus.* Drain. Pierce segments with a fork before placing on trays. Evaporation will be speedier. Remove seeds if there are any. Dry at a low temperature of 140° F. Store in airtight containers when dry.

Or 3 cups of segmented fruit may be placed into a simmering syrup solution consisting of 4 cups sugar to 4 cups water. Simmer fruit for 3 minutes. Let cool in syrup 10 minutes, then drain and place on trays. Pierce fruit with a fork and remove seeds if there are any. Dry at a low temperature of 140° F. Store in airtight containers when dry.

LEMONS. –

The lemon is one of the most versatile fruits. Lemons rank high as an excellent source of vitamin C but their use in foods, hot or cold drinks, and as a cleansing aid make them the handyman of the kitchen.

Some varieties of lemons are more tart than others but the Meyer lemon in particular is noted for its sweet taste. This golden fruit is esteemed by the

Mexicans and it's not unusual to see the children eating lemons with as much gusto as the American children eat sweets.

Top-grade lemons have a fine-textured, oily, elastic, bright yellow skin with pointed ends. Also check the weight of the lemons; the heaviest fruit contains the most juice. Large-knobbed lemons denote less juice. Fruit should be snipped off with sharp shears and then handled very carefully to avoid bruising.

Lemons should be kept chilled and any cut portion should be carefully wrapped. Lemons that are dry will become juicier and can be rejuvenated by placing in an oven and heating through. One medium-sized lemon will produce 3 tablespoons of juice.

All parts of the lemon are indispensable. Even the culls are used in manufacturing citric acid, lemon juice, and lemon oil, and in making marmalades. The spicy oil of the rind is used in extracts and candy. The white spongy portion of the peel contains pectin, the basis of jellies. Instructions for making pectin are on page 153.

Lemon juice perks up appetites, calms and soothes nerves. Lettuce and other salad greens will revive and become crisp if lemon juice is added to the water in which they are washed, as it acts as a freshener. Lemon juice makes a refreshing hot or cold drink. As a garnish, it adds a festive touch to any dish. A generous amount of lemon juice adds zest to salads. Fish and meats are enhanced by its flavor and fruits and vegetables are delicious when it is used as a natural dressing.

Dull or blackened aluminum can be brightened by cleaning with a cloth dipped in lemon juice. A small amount on hands removes fish and onion odors. It will also keep the hands soft and white. A combination of 1 teaspoon lemon juice and ¼ teaspoon baking soda makes a simple remedy for skin blemishes.

To dry: Segmented lemons may be processed two ways. One way is to place segments into sodium bisulfite solution or another desired solution. Follow the instructions given in the section *Pretreatment of Foods* and the heading *Citrus*. Drain. Pierce segments with a fork before placing on trays. Evaporation will be speedier. Remove seeds if there are any. Dry at a low temperature of 140° F. Store in airtight containers when dry.

Or 3 cups of segmented fruit may be placed into a simmering syrup solution consisting of 4 cups sugar to 4 cups water. Simmer fruit for 3 minutes. Let cool in syrup for 10 minutes, then drain and place on trays. Pierce fruit with a fork and remove seeds if there are any. Dry at a low temperature of 140° F. Store in airtight containers when dry.

LIMES. —

Limes are from a small spiny tropical tree of the rue family. The trees are raised for their sharp acidy fruit. Limes are usually only half the size of lemons and almost round in shape. They are very strong in citric acid and in years past were carried aboard ship to prevent scurvy. This citrus has a bright green color that contains a juicy but very acrid pulp.

There are two kinds of limes, the Persian and the Dominican. The Persian is seedless and the larger of the two. As in all citrus, the best fruits are bright in color and heavy in size. Deep-yellowed limes do not have the desired acidity.

Limes are used in the same manner as one would use lemons. They add zest and tang in whatever they are in. They perish quickly and should be carefully washed and chilled until needed.

To dry: There are two ways to dry limes. One way is to place segments into sodium bisulfite solution or another desired solution. Follow the instructions given in the section *Pretreatment of Foods* and the heading *Citrus*. Drain. Pierce segments with a fork before placing on trays. Evaporation will be speedier. Remove seeds if there are any. Dry at a low temperature of 140° F. Store in airtight containers when dry.

Or 3 cups of segmented fruits may be placed into a simmering syrup solution consisting of 4 cups sugar to 4 cups water. Simmer fruit for 3 minutes. Let cool in syrup solution for 10 minutes, then drain and place on trays. Pierce fruit with a fork and remove seeds if there are any. Dry at a low temperature of 140° F. Store in airtight containers when dry.

ORANGES. —

The orange is a botanical plant and fruit of the genus *Citrus*, a member of the rue family. Orange leaves are glossy and their flowers are a "bit of heaven" they are so delicately scented. This citrus is widely used in diets as they contain vitamins A, B, and C as well as needed minerals.

Choice quality oranges are firm, well formed, and heavy for their size. The skin should be smooth, thin, and free from disease and decay. They should be sweet and juicy. Avoid puffy oranges as they lack juice and are usually of a poorer quality.

The juice of the orange contains sugar and citric acid while the white, spongy part of the peel contains pectin. This substance is used in preserves, jams, jellies, and marmalades. To process pectin follow the instruc-

34

tions given for *Pectin* on page 153.

To dry: There are two ways to process oranges. One way is to place segments into sodium bisulfite solution or another desired solution. Follow the instructions given under the section *Pretreatment of Foods* and the heading *Citrus*. Drain. Pierce segments with a fork before placing on trays. Evaporation will be speedier. Remove seeds if there are any. Dry at a low temperature of 140° F. Store in airtight containers when dry.

Or 3 cups of segmented fruit may be placed into a simmering syrup solution consisting of 4 cups sugar to 4 cups water. Simmer fruit for 3 minutes. Let cool in syrup solution for 10 minutes. Drain and place on trays. Pierce fruit with a fork and remove seeds if there are any. Dry at a low temperature of 140° F. Store in airtight containers when dry.

TANGERINES. —

Tangerines are a variety of the mandarin group. Their color is similar to mandarin orange. They resemble the orange somewhat, but are flatter at each end. Tangerines contain a very sweet juice, a loose, easily detachable rind, and the fruit itself is easily segmented. The thin peel has an aromatic scent. Add a bit of dried peel to a cup of tea and welcome its delicate spicy flavor. It is especially delightful on a wintery, blustery day.

To dry: There are two ways to process tangerines. One way is to place segments into sodium bisulfite solution or another desired solution. Follow the instructions given in the section *Pretreatment of Foods* and the heading *Citrus*. Drain. Pierce segments with a fork before putting them on trays. Evaporation will be speedier. Remove seeds if there are any. Dry at a low temperature of 140° F. Store in airtight containers when dry.

Or 3 cups of segmented fruit may be placed into a simmering syrup solution consisting of 4 cups sugar to 4 cups water. Simmer fruit for 3 minutes. Let cool in syrup solution for 10 minutes; then drain and place on trays. Pierce fruit with a fork and remove seeds if there are any. Dry at a low temperature of 140° F. Store in airtight containers when dry.

CITRUS PEELS.—

Cut, slice and peel into desired pieces and steam for 10 minutes to remove bitter taste. If you use sodium bisulfite solution the peel will retain its bright color. Drain and place on trays in a low oven. Keep the oven door completely open. Store in airtight containers when dry. Citrus peel is excellent when served with tea. Use to season or spice food. It can be ground if desired.

Or cut into desired slices or pieces and combine with 4 cups water to which ¼ teaspoon of baking soda has been added. Simmer covered for 20 minutes stirring occasionally. Rinse. Drain and place on trays in a low oven to dry. Store in airtight containers when dry. The peel will be less fragrant but less biting with this method.

Cranberries

The cranberry belongs to the heath family. It has small shiny leaves and dark lustrous red berries that vary in size and color. Cranberries should be picked as quickly as possible as they soon become overripe. Wash well just before using.

Chop or leave whole. Place chopped cranberries on trays. Whole cranberries may be steam blanched for 2 minutes to crack the skins. Place on the trays in a low oven at 140° F. until dry. Paper toweling may be placed under the trays and aluminum foil may be placed on the bottom of oven. Watch paper toweling carefully to see that it doesn't burn. Remove paper toweling when fruit firms up. The oven door on a gas range must be propped open 8 inches, on an electric range half an inch. Do not overdry.

Drying time: 2-4 hours.

To reconstitute: Add 1 cup cranberries to 2 cups of water. They can be soaked for 30 minutes, then simmered for several minutes or until tender. Add seasoning if desired.

Dates

Dates that have reached their prime condition are ripe, smooth-skinned, plump, and a lustrous golden brown. To cut dates into cubes, place pitted dates on a chopping board and cut with a sharp knife or use kitchen shears. Dip knife or shears frequently into cold water to prevent fruit from sticking. Place in desired solution. See the section *Pretreatment of Foods*. Drain. Place on trays in a low oven at 120° F. The oven door on a gas range must be propped open 8 inches, on an electric range half an inch. Dry to desired firmness (check a cool sample). Store in airtight containers in a cool place when dry.

Dates are one of the few dried fruits that do not need to be reconstituted.

Figs

Fig trees do not have outer blossoms as the flowers

are inside the pear-shaped fruit and this is why they are so full of seeds. Ripe figs should be firm but soft to the touch. The color of the fruit will depend on the variety. Partially ripened fruit will result in a tasteless dried product. Dropped fruit should be picked up immediately to avoid insect infestation. Figs also sour and ferment quickly and the odor is very noticeable when this occurs. For this reason, figs should be refrigerated.

Fruit may be cut in half or left whole. No prior treatment is necessary although they may be treated in a variety of ways:

1. They may be treated with one of the solutions in the section *Pretreatment of Foods.* - pg. 10 + 11

2. They may be steamed for 20 minutes.

3. Or they may be dropped into a simple syrup (1 cup of sugar to 3 cups water) and simmered for 10 minutes. Remove from heat and let stand 20 minutes.

Drain. Place on trays in a 120° F. oven, raising temperature gradually to 145° F. until dry. The oven door on a gas range must be propped open 8 inches, on an electric range half an inch. Turn fruit to prevent sticking. The dried fruit should be pliable and somewhat sticky with a dry center. Store in airtight containers when dry.

Drying time: 4-7 hours.

To reconstitute: Dried figs should only be soaked a short time and cooked quickly to return them to their former state. Add 1 tablespoon of sugar to each cup of fruit the last few minutes of cooking. Dried figs may also be steamed for 20 minutes or until plump. They will also blend well with other ingredients if they are soaked for 5 minutes in a small amount of boiling water.

To perk up their bland flavor, add grated citrus peel or any of the following spices: cinnamon, cloves, or candied ginger.

Fruit Leathers

This is one of the simplest ways of preparing delightful, nutritious treats. It is also an excellent way of using overripe fruit. Fruit leathers are lightweight and can be easily carried on trips. They add more nutrition and taste to other foods. They can be eaten as a dessert or as a snack for in between munching. Fruit leathers keep almost indefinitely if prepared and packaged correctly and are very, very easy to make.

Any fruit can be used. Peeling is optional but certain fruits do not have eye appeal when skins are blended with the pulp. Cover with water and cook until soft. Sieve, strain, or place in blender. Puree must be the consistency of applesauce.

Combinations of mixed fruits are an added treat. For taste variety, add spices, grated peels, crushed seeds, and crushed nuts. Leathers may be sweetened with sugar, syrup, or an artificial sweetener. They may also be sweetened with honey, but the puree must be warmed first. Remember if it tastes good after it has been blended and sweetened it will taste good as a snack later. Some fruits do not need a sweetener, while others call for 1-2 tablespoons per cup. Sweeten to taste. If using an artificial sweetener, follow the instructions given on the package.

Line a cookie sheet with plastic wrap. Pour 2 cups puree over the surface, taking care not to spread the fruit to the edges as the moisture might seep under.

Edges are also necessary for easy removal of the fruit leather later. Purees should be dried at the lowest possible oven temperature. The oven door on a gas range must be propped open 8 inches, on an electric range half an inch. The oven door may be completely open toward the end of drying time. When leather is firm, like plastic, and can be removed easily, then it is ready to be packaged. When cool, remove plastic film and roll carefully. Cut into desired pieces and store in airtight containers. Check periodically for any problems.

Remember fruit leather is concentrated and adds calories to the diet.

Grapes (Raisins)

Raisins are grapes that have been dried with the sun's warmth or through other means of dehydration. Varieties of grapes native to America will grow in almost any region, from the temperate to the subtropical; while the European grapes will grow only in Arizona and California. Thompson, Muscat, and Alexandria grapes are the three types usually favored for raisins, with the Thompson considered the superior of the three. Other flavorful grapes may also be dried but their skins are tough and fruit must be halved, quartered, or ground before using in recipes.

Prime quality grapes are fresh and usually plump with the fruit very firmly attached to the stems. An excellent color denotes well-developed flavor and a high sugar content. The Thompson grapes that have turned amber are especially delightful—as sweet as honey and extremely flavorful. They are also seedless and do not take as long to dry as the varieties with seeds. Grapes with seeds have a stronger flavor and the seeds must be removed. This is easily accomplished by pressing gently when grapes are partially dried. Muscat grapes are processed in the same way as the Alexandria grapes.

Grapes must be picked with care as they are easily crushed and bruised if roughly handled. Avoid damp grapes as they will mold and decay easily. Grapes should be washed, stemmed, and blanched in the desired solution. See the section *Pretreatment of Foods.* Blanching is necessary to cleanse fruit but more important, blanching cracks the skins so that the inner moisture can evaporate during the drying process. Otherwise fruit would remain moist in the center. Treated grapes will be lighter in color when dry.

Place on trays in a 140° F. oven with the door on a gas range open 8 inches, on an electric range half an inch. Paper toweling may be placed under trays and aluminum foil at the oven bottom to catch spills. Watch carefully that paper doesn't burn. Paper toweling may be removed when food firms up. Store in airtight containers when dry. Raisins may be eaten dry or used in your favorite recipe.

Drying time: 3-5 hours.

To reconstitute: raisins should be soaked in boiling water 5 minutes before using.

Gourmet treat: Pick grape clusters when fully ripe. Do not wash but carefully *snip* off any grapes that are spoiled or immature. Place grapes in a crock, stem side up. Cover grapes with a solution of molasses and water—adding just enough molasses to color slightly. Be sure grapes are fully immersed in the liquid. Cover with a lid or dish to hold the fruit under. Place in a dark, cool place for 4 months. Remove the fruit as needed and soak in cold water for a short period of time. These delicious grapes will have a hint of warmth and a sharp sprightly taste.

Guavas

Guava is a genus of the myrtle family, with pulpy fruit about the size of a small apple, containing many hard seeds. Colors vary from yellow to red and its shape ranges from the ovate to the round. It is popular in tropical countries as a fresh or delicious stewed product and can be used in any recipe calling for papayas or citrus fruits.

Remove seeds and center pulp. Slice fruit into desired pieces with a stainless steel knife to prevent discoloration. Place in desired solution. See the section *Pretreatment of Foods*. Drain. Place on trays in a 160° F. oven for several hours, lowering temperature to 140° F. until dry. Keep the oven door on a gas range open 8 inches, on an electric range half an inch. Store in airtight containers when dry.

Dried product should contain 7 percent moisture.

Nectarines

Nectarines are a variety of peach with a smooth, waxy skin and a firm but very fragrant flesh. There are freestone and clingstone nectarines as there are in peaches. Fruit that is not very ripe should be left to mellow in a warm room and then refrigerated. Avoid green fruit as it will wither or shrivel before it ripens.

To dry: Nectarines do not have to be peeled but are more attractive when peeled. They may be steamed 5-10 minutes but this is unnecessary for a good dried product. Nectarines do not have to be sulfured. A little caution in working with small amounts to prevent darkening will yield an excellent product that is extremely tasty with considerable eye appeal. Cut into halves, slices, or bite-size chunks. Drop into desired solution to prevent fruit from darkening. See the section *Pretreatment of Foods*.

Place on trays in a 140° F. oven with the oven door on a gas range propped open 8 inches, on an electric range half an inch. You may place paper toweling under trays and aluminum foil at the oven bottom. Watch paper carefully so that it doesn't burn. Remove paper toweling when fruit firms up. Store in airtight containers when dry.

Drying time: 7-14 hours for halves.

To reconstitute: For fresh nectarines use 1 cup nectarines to 1 cup water. Chill in the refrigerator. Add ginger ale or a touch of candied ginger before serving.

For stewed nectarines use 1 cup nectarines to 2 cups water. Simmer until tender, adding seasonings at the end of the cooking period.

Papayas

Papaya is the fruit produced by the papaw tree that grows in the Southern and Central United States. Its fruit is tangy and slightly sweet. It can eaten fresh or served with salt, pepper, sugar, or lemon. The seeds can also be eaten and are considered a delicacy.

Papayas are shipped green and are usually bought in this state and allowed to ripen. The semiripe fruit can be used with sugar for preserves or it may be cooked and used as a vegetable. When ripe, it serves as a delightful treat in a salad or dessert. Green papayas also contain a substance, papain, that is used to tenderize tough cuts of meat.

Peel, deseed, and slice fruit into ½ inch pieces. Dip into desired solution. See the section *Pretreatment of Foods*. Drain. Place on trays in a low oven until dry. You may place paper toweling under tray and aluminum foil at the oven bottom. Watch paper toweling carefully so it doesn't burn. Remove paper toweling when fruit firms up. The oven door on a gas range must be propped open 8 inches, on an electric range half an inch. Store in airtight containers when dry.

To reconstitute: Cover with water (add more later if necessary). Place in refrigerator to crisp. Papayas can be served fresh in cookies, cakes, etc. or can be eaten dried. They make a good snack treat.

Peaches

There is no accurate account of where the peach tree originated. Records, however, go back many centuries before the time of Christ. Some authorities claim they came from China. It is known that the Greeks of Alexandria used the fruit for feasting as well as an ornament.

Botanically it belongs to the drupe family but there is controversy over whether it should be listed along with the plums, cherries, and apricots.

There are many varieties of clingstone and freestone peaches, but the freestones are the easiest to work with. Regardless of the type you use, the peach should be fresh, firm, and brightly colored from yellow to almost white at times, with a blush of rose. Fruit should be picked when flesh gives slightly under thumb pressure. Fruit must be mature, juicy, and sweet to produce a good dried product.

Try rubbing off as much fuzz as possible. Peaches do not have to be peeled but are more attractive when peeled. They may be steamed 5-10 minutes but this is unnecessary for a good dried product. Peaches do not have to be sulfured. Working quickly with small amounts and handling them carefully will prevent darkening and will yield an excellent product that is extremely tasty with considerable eye appeal. Cut into halves, slices, or bite-size chunks. Drop into desired solution to prevent fruit from darkening. See the section *Pretreatment of Foods*.

Place on trays in a 140° F. oven with the oven door on a gas range open 8 inches, on an electric range half an inch. You may place paper toweling under trays and aluminum foil at the oven bottom. Watch carefully that paper doesn't burn. Remove paper toweling when fruit firms up. Store in airtight containers when dry.

Drying time: 5-7 hours for halves.

To reconstitute: For fresh peaches use 1 cup peaches to 1 cup water. Chill in the refrigerator. Add ginger ale or a touch of candied ginger before serving.

For stewed peaches use 1 cup peaches to 2 cups water. Simmer until tender, adding seasonings at the end of the cooking period.

Pears

The pear is one of the easiest fruits to dry and one of the tastiest. Any ripe pear may be used but the Bartlett pear is one of the best. The flesh is sweet and juicy and is excellent for home use, canning, freezing, and drying. Pears can be picked while still green (not too green) and will ripen gradually, becoming sweeter and juicier as they mature. Pears that are left on the tree to ripen are often coarse textured and in some varieties grainy.

Mature pears yield to pressure applied at the stem's base and have a flush of yellow, at times deepening into a rose. Their bland flavor easily blends with other more highly flavored fruits and will accentuate the taste when mixed with sharp-flavored foods.

Pears may be dried with the skin on, but peeled pears are less grainy. (Peelings may be simmered and strained for sauce or restrained for fruit leather.) Pears should be placed into a solution. See the section *Pretreatment of Foods.* This treatment will result in dried fruit that will be tasty, delicious, and light in color.

Peel, core, and section fruit into bite-size chunks, slices, or dices and drop into preferred solution. Drain and place on trays in an oven heated to 160° F. with the oven door on a gas range open 8 inches, on an electric range half an inch. You may place paper toweling under trays and aluminum foil at the oven bottom to catch drips. Watch carefully that paper doesn't burn. Remove paper toweling when fruit firms up. Remove fruit from oven and allow to cool before checking for dryness. Dried fruit should be pliable but leathery. Store in airtight containers when dry. This makes a delicious snack food.

Drying time: 5-7 hours for 1/8 inch pieces.

To reconstitute: Cover with water and chill for fresh fruit. Or use 1 cup dehydrated pears to 2 cups water and simmer for 20 minutes until tender. Add flavorings and sugar during the last few minutes of cooking.

Persimmons

Persimmon is a common name for a genus of small trees of the ebony family. <u>Persimmons should be eaten when fully ripe as they are very astringent when green due to the large amount of tannin present</u>. If not too green, this taste can be neutralized by freezing, by placing into warm water for a long period of time, or by adding ¼ teaspoon of baking soda to the pulp before heating. Persimmons will discolor if cooked in tin vessels.

They should be washed, chilled, and served with a grapefruit spoon so that the delicious, flavored fruit may be easily removed. Add wine or a kiss of lemon or lime for a gourmet touch. Add persimmon pulp to salads, sauces, ice cream, and jellies.

To dry: Wash, cut fruit in half, and place into desired solution. See section *Pretreatment of Foods.* pg 10+ 11 Drain. Place on trays in an oven heated to 140° F., until dry. You may place paper toweling under trays and aluminum foil at oven bottom to catch drips. Watch carefully that paper doesn't burn. Remove paper toweling when fruit firms up. The oven door on a gas range must be propped open 8 inches, on an electric range half an inch. Store in airtight containers when dry.

Drying time: 5-7 hours for halves.

To reconstitute: Immerse in water to cover. Place covered into the refrigerator to chill for several hours. They are best eaten as a snack food.

Pineapples

Pineapples were discovered in the fall of 1493 by Columbus on his second voyage westward. Hundreds of years later they were found in Hawaii although they are not native to that area. Today, Hawaii is pineapple heaven and pineapple growing and processing is one of the principal industries of the state.

Pineapples are very sturdy and can take rough handling if they are harvested green and allowed to ripen on the way to the market. Field-ripened pineapples, however, must be picked and canned within 24 hours to prevent spoilage. Sun-ripened pineapples are superior to those bought in the market.

The quality of pineapples can be determined by their appearance, orange yellow color, fragrant odor, heavy weight, and leaves that pull away easily. The yellow fruit is delicious and tangy and can be eaten fresh, served as appetizers, used in salads, or as a festive and decorative touch to chicken and meats. A serving of pineapple after heavy dining is a light but refreshing treat.

Pineapples are easily pared by removing the top spikes with a sharp knife. Work over a low bowl to catch drippings. Peel, then cut out the eyes. The trimmings are sour and should never be eaten. However, they can be covered with water and allowed to stand for a few hours. Drain off the liquid, chill, and serve with soda water over crushed ice.

Prepared fruit may be sliced, diced, or chopped. Place into desired solution. See the section *Pretreatment of Foods*. Place on trays in a low oven with the oven door on a gas range propped open 8 inches, on an electric range half an inch. You may place paper toweling under trays and aluminum foil at the oven bottom to catch drips. Watch carefully that paper doesn't burn. Paper toweling may be removed when food firms up. Store in airtight containers when dry. These are delicious eaten in their dried state.

To reconstitute: Add water to cover and chill for several hours. Fruit will be crisp.

Plums or Prunes

Plums and prunes are drupes or true stone fruits, of the genus *Prunus*. The trees are cultivated for their smooth-skinned fruits. Prune plums are excellent for drying but regular plums may be dried too. They take longer to dry and must be pitted and halved. The prune plum may be dried whole or halved. Whole prunes may be blanched in steam until the skins crack, or simply cover the firm, ripe fruit with boiling water and let stand 10 minutes. Bisulfiting will speed up the drying process and preserve the fruit in flavor and color.

Prunes are a wonderful source of natural energy and in their prime state are firm and sweet. Softening at the tip is a good indication of maturity. Overripe fruit should be reserved for desserts, jams, jellies, and fruit leathers. Avoid fruit that is soft and runny as it is usually tasteless in flavor. If purchased, stains on the container will indicate this overripeness or decay.

Wash, pit, and dip into desired solution. See the section *Pretreatment of Foods*. Place on trays, cup side up, in an oven heated to 140° F., gradually increasing temperature to 160° F. until dry. The oven door on a gas range must be propped open 8 inches, on an electric range half an inch. You may place paper toweling under trays and aluminum foil at the oven bottom to catch drips. Watch carefully that paper doesn't burn. Toweling may be removed when food firms up. Dryness is indicated when fruit is firm to the touch and the pit cannot be separated from the flesh by rolling prune between the thumb and two fingers. Fruit that is underdried will mold. If this occurs, destroy product.

Drying time: 4-6 hours for halves.

To reconstitute: Add 1 cup prunes to 1½ cups water. A slice of lemon is optional. Cook prunes slowly until tender and juicy. Do not boil. Sugar may then be added if desired. Or prunes may be covered with boiling water, chilled, and refrigerated for 24 hours.

Prunes are delicious when served with milk or cream. Allow to stand for several hours until the milk absorbs the sweetness. Prune fillings are also delicious and the fruit is equally as tasty in desserts, whips, and sauces.

Raisins

See *Grapes*.

Rhubarb

A thick, hardy perennial herb which is often regarded as a fruit but is really a vegetable. It is sometimes called the pie plant. Its delicious tart sauce is eagerly welcomed in the spring. Rhubarb contains vitamin C and may be used as an excellent substitute for citrus fruits. It may be used in sauces, pies, jams, and preserves. It is also used at times for medicinal purposes—as a mild laxative. Rhubarb is colorful and extremely zesty. Its tenderness and firmness can be determined by easy puncturing of the stalks. Do not eat the leaves as they contain oxalic acid salts.

Use only tender young stems to dry, as the older stalks are usually tough and stringy. Wash, trim, and cut into desired pieces. Steam blanch about 2-4 minutes. Overblanching will cause rhubarb to mat. Drain and place on trays. You may place paper toweling under trays and aluminum foil at the oven bottom to catch drips. Watch carefully that paper doesn't burn. Paper toweling may be removed when food firms up. Store in airtight containers when dry.

To reconstitute: Place in water to cover for several hours.

Watermelon

Watermelons are an annual trailing vine of the cucumber family, cultivated for their round or oblong green, mottled-striped fruit. Their pink or red pulp is sweet and refreshing. Watermelon is an elegant summer treat when chilled and served on a hot day. Its refreshing coolness complements fruit salads and desserts. Pickled watermelon rind is extremely tangy and zesty and makes an appetizing relish or condiment. Add a few maraschino cherries to your favorite recipe. Sometimes the watermelon seeds are crushed and steeped as a tea for treating kidney trouble.

Watermelons should be left on the vine until ripe. The gradual yellowing and hardening of the underside of the fruit is one indication of full maturity. The ripe melon should have a dull appearance and there should be a definite dull thumping sound when flicked with the finger. Immature melons have a hard, green, unripe appearance, while the overripe ones have a dull coat and are soft and springy to the touch.

To dry: Remove rind. Cut fruit into ½ inch slices or smaller pieces and place into desired solution. See the section *Pretreatment* (on page 41) *of Foods*. Allow fruit to drain for 30 minutes, then place on trays in an oven at 140° F. The oven door on a gas range must be propped open 8 inches, on an electric range half an inch. You may place paper toweling under trays and aluminum foil at the oven bottom to catch drips. Watch carefully that paper doesn't burn. Paper toweling may be removed when food firms up. Drying time is longer as watermelons have a high water content. Finish drying at a low temperature or leave the oven door completely open. Store in airtight containers when dry.

Drying time: Seems like forever!

To reconstitute: Place in water to cover. Watermelon chips make a better snack food and are tastier than the reconstituted fruit. When dry, watermelons have a small yield as the fruit is mostly composed of water.

chpt. 6.

gro, fm pg 53 to 101 = 49 pgs

Drying Vegetables

The quality of the dried vegetable depends on many factors: the type of vegetable, its freshness, the method of handling, the promptness of preparation, your care in drying, and its proper packaging. Followed through, the result will be a durable product, a minimal loss of nutritive values, and a finer reconstituted vegetable. Carelessness in preparation will result in an inferior vegetable or partial or even total loss of the product.

Many vegetables are best in their immature state: when the flesh is tender and succulent, needing only a small amount of cooking to bring their natural goodness to the table. Some vegetables are merely brushed or sautéed with butter and seasoned with a judicious touch of cheese to bring out or add to their wonderful flavor. Other vegetables are in their prime when fully mature. Vegetables range in taste from mild to zesty, yet all vegetables add sparkle and spice to a meal.

Home gardening, with its many advantages, is one answer to vegetable drying. Foods can be picked at their peak in quality and flavor. One can grow unusual vegetables not found in the local markets, better varieties can be raised, and a considerable savings can be realized by having a garden patch. Gardens bring in excellent food for the table with the excess going into the oven for drying. Gardening can turn into a wonderful and profitable hobby.

If home-grown vegetables are unavailable, market produce is the answer. You will find these vegetables rougher and coarser in texture—a necessary quality to withstand handling and shipping. Buying produce in bulk greatly reduces the price of the products. Remember that freshness is important in some vegetables and unnecessary in others. (Peas and beans must be fresh as opposed to onions and potatoes.) Check for blemishes and defects. Wilted or poorly developed plants are a waste of time, energy, and money.

Because the handling and preparation is attended to at home, one can work with ease and with quantities that are easy to handle. Vegetables that are perishable and cannot be dried quickly should be refrigerated promptly. Carefully examine and discard any old and yellow leaves or diseased vegetables. Some vegetables are susceptible to insect infestation and should be soaked in a mild salt solution for about 20 minutes to drive out the insects. Some are washed, peeled, or cut to ready them for blanching or drying. Quick handling is important as foods may spoil or lose their excellent qualities. All equipment should be kept clean, including knives, utensils, bowls and the working area. This will deter bacteria. Remember to remove any foods that smell or look unwholesome. One small particle of spoiled food may contaminate the whole batch.

Pretreating foods is a necessary step in drying foods.

See the section *Pretreatment of Foods*. (pg 10 + 11) Bisulfiting cuts down on drying time, retains the original color, and the product takes less time to reconstitute. It should not be used to treat legumes as it destroys some of the nutrient value. Salt and vinegar solutions are also effective in maintaining color. Blanching and steaming are two other very important steps in drying. These methods preserve the color and vitamins, hasten drying time by relaxing the cell structure and tissues, and the fresh flavor of vegetables will be maintained when reconstituted later. Unblanched or unsteamed vegetables will not maintain their color and do not reconstitute well. If steaming is not possible, the vegetables should be placed into simmering water for a few minutes. Do not boil. However, with this method there is a loss of vitamins and minerals. Avoid adding too many vegetables to the water at one time as this will lower the water temperature.

Steaming varies with the vegetable. Leafy greens and celery will steam in 3-4 minutes while other vegetables will take anywhere from 6 to 30 minutes. It is very important that the vegetables are tender and firm. Beans, potatoes, peas, and spinach must be rinsed after blanching. Do not combine strong-flavored foods with mild-flavored foods when drying unless they are being prepared for use in soups.

Vegetables are dried at different temperatures. The oven door on a gas range must be propped open 8 inches, on an electric range half an inch. The temperature on all foods must be lowered toward the end of drying time. If this is not possible, leave the oven door completely open until the food is dry, removing any food that is ready for packaging. Trays should be rotated and dried food removed. Refrain from adding fresh foods to partially dried foods as the partially dried foods will absorb the moisture of the fresh foods and prolong drying time overall. An oven temperature that is too low will sour food while a temperature that is too high will form a seal and prevent moisture from escaping. Good air circulation is necessary so do not overload the trays. Be sure that the trays are at least 3 inches above the oven bottom. Foods ahould always be cooled before checking for dryness.

Package all dried foods carefully. Be sure they are in moistureproof bags or containers. They should be stored in a cool dark place. Check periodically for infestation or mold. If molding should occur, destroy the product.

All dried vegetables contain starches and should be thoroughly cooked. Starches, when not thoroughly chewed, may cause digestive disturbances.

There are many ways that dried vegetables may be used. Try cooking them in meat or chicken stock for a slightly different but tasty dish. Combine a variety of vegetables and blend into a powder for soups. These are delicious on a cold and blustery day. Minimal cooking will produce a hearty and warming soup. Vegetables and cooked meats may be blended to make a crunchy, munchy cracker. This mixture may also be poured over a layer of oatmeal and baked for another delicious and appetizing treat.

Remember that some vegetables are saltier or sweeter when reconstituted and for this reason seasoning should be withheld until the end of the cooking period. Seasoning also retards reconstitution.

① Artichokes

The French or globe artichokes have dark green or purplish heads. They grow to 4 or 5 inches in diameter, although the smaller heads are more tender. Use only the tender hearts. Trim the stem and cut in half. For better color, the artichokes may be placed into sodium bisulfite for a second or placed into another desired solution. See the section *Pretreatment of Foods*. Steam blanch for about 15 minutes. Place on trays in an oven set at 140° F. for several hours. The oven door on a gas range must be propped open 8 inches, on an electric range half an inch. Lower the temperature near the end of the drying time. Package carefully in airtight containers when dry.

Drying time: 4-6 hours for 1/8 inch strips.

To reconstitute: Add 1 cup artichoke to 1 cup water. Use in recipes calling for fresh or frozen artichoke hearts.

See also *Jerusalem Artichokes*.

② Asparagus

The appearance of asparagus is one of the first signs of spring and although the season is short, with modern methods it is now available throughout the year.

For eating, dry the tips only, using young tender stalks. For seasoning and cream of asparagus soup, dry the older stalks and powder in blender.

Wash asparagus carefully and cut off all woody stems. Steam blanch for 5 minutes, then dry. If desired, they may be sulfited before steaming. See the section *Pretreatment of Foods*. Place on trays in the oven at 140° F., allowing the air to circulate freely. The oven door on a gas range must be propped open 8 inches, on an electric range half an inch. Turn the stalks often.

Drying time: 3-6 hours.

To reconstitute: Add 1 cup dried asparagus to 2 cups unsalted water. Soak for 30 minutes. Cover, bring to a boil. Lower temperature and simmer for 30 minutes or until done. Add salt during the last stages of cooking.

(3) *Beans*

There are a variety of beans ranging from the edible pods on green and wax beans that can be eaten in their entirety to the matured seeds of different varieties. The outer pods of baby limas, kidney, navy, and pinto beans should have a fresh appearance and when shelled, beans should be plump and light colored. They should be blanched in boiling water for 5 minutes or steamed for 10 minutes.

Green and wax beans (yellow) should be picked when young and tender and so crisp that they snap easily. All the beans should be at the same stage of ripeness so they will cook evenly. Remove strings if any, wash, cut into desired lengths or slice lengthwise. Beans may lose some of their nutrient value if dipped into sodium bisulfite solution, so one should choose another desired solution. See the section *Pretreatment of Foods*. They may also be steam blanched for 10-15 minutes or until tender. Green beans should be rinsed in cold water after blanching.

To dry: Drain, place on trays in an oven set at

115° F., gradually increasing heat to 140° F. The oven door on a gas range must be propped open 8 inches, on an electric range half an inch. Package carefully in airtight containers when dry.

Drying time: 6-10 hours for short cut green and wax beans.

To reconstitute: Dried baby limas, kidney, navy, or pinto beans should be soaked overnight before using. Cover with water ½ inch above the beans. Simmer slowly until tender and done. Season toward the end of cooking time.

Dried green or wax beans should be soaked in water to cover for 30 minutes and then simmered for 15 minutes or until tender. Add seasonings the last few minutes of cooking time.

Beets

The late beet crop is excellent for drying as there is more uniformity of color than in the spring crop. Beets are globular or top-shaped. They should be free from blemishes and zoning (white bands). Immature beets should be avoided. Beets should be solid, dark red, or reddish purple. The medium-sized beet is excellent for drying and when dried, color deepens to purplish-black.

BEET LEAVES. —

Young leaves are tender and have a mild but excellent flavor. Older greens are inclined to be strong. Remove stems and discard yellow or wilted or damaged leaves. Leaves must be cleansed thoroughly to remove any insect spray, soil, or dirt. Change the rinse water several times. Try to remove as much water as possible, patting dry with paper towels. Care should be maintained throughout the handling process to prevent bruising and crushing. Beet leaves must be processed quickly to prevent loss of vitamins. Beet leaves should be processed within 5 hours. Excess greens should be held under refrigeration and dried as quickly as possible to insure vitamin content, good taste, and flavor.

Leaves should be dried whole in order to lessen vitamin loss through bleeding. Beet leaves should be steamed and placed on trays as quickly as possible. Blanching time varies from 2 to 4 minutes or until the ribs are translucent. Overblanching will cause the soggy leaves to mat and stick to the trays and drying time will be extended.

Beet leaves may be sulfited or you may use another desired solution. See the section *Pretreatment of Foods*. Pretreatment helps the succulent leaves hold their color and flavor, deters vitamin loss, and reduces drying time appreciably. The reconstituted product will have more taste and eye appeal. Drain well, spread food evenly on trays and into prescribed temperature. Avoid overcrowding.

Oven temperature should not exceed 100° F. As the product nears completion, care should be taken that the leaves do not brown. Oven door may be left open and contents frequently supervised to prevent browning. When dry, store in airtight containers. Do not crush leaves.

To reconstitute: Add 1 cup dried beet leaves to 1 cup boiling water. Simmer gently for 2-3 minutes. Do not overcook.

60

⑥ BEET ROOTS. —

The beet tops and roots should be trimmed about an inch. Beets should not be peeled before cooking—they lose their color. To prevent bleeding, a minimum of water should be used. Cook for 25 minutes or longer, depending on the size of the beet. Or steam whole for 30 minutes or until tender. Long cooking may cause deterioration of flavor and color. Cool. Slip off skins and cut into desired pieces, slices, cubes, or julienne strips. Place on trays in an oven at 140° F. The oven door on a gas range must be propped open 8 inches, on an electric range half an inch. Store in airtight containers when dry.

Drying time: 3-6 hours for 1/8 inch slices.

To reconstitute: Add water barely to cover and soak overnight or soak for 30 minutes. Bring to a boil and simmer gently 10 minutes or longer until tender.

⑦ Broccoli

Broccoli, a type of cauliflower, belongs to the same family as the cabbage and is cultivated for its tender underdeveloped flowers. The flowers should be tightly closed as the broccoli is past its prime if flowers are open or yellowing.

Broccoli should be washed carefully and the woody stems trimmed as in asparagus. The stems of older broccoli should be peeled thinly. Flower heads should be soaked in a salt solution for 30 minutes to check for infestations. Cut into desired pieces. Steam broccoli for 10 minutes or water blanch for 5 minutes. Broccoli can be dipped in sodium bisulfite, or another desired solution. See the section *Pretreatment of*

Foods. Process stems and heads separately or start drying the stems before the heads.

Place on trays in an oven set at 140° F. for 3 hours or longer until dry. Keep the oven door on a gas range open 8 inches, on an electric range half an inch. Broccoli should be brittle or crisp. Store in airtight containers when dry.

Drying time: 3 or more hours.

To reconstitute: Add water to cover and let stand 30 minutes. Gently simmer until tender.

⑧ Brussels Sprouts

Brussels sprouts are miniature cabbages with somewhat globular compact heads. They are one of the many vegetables developed from the original wild cabbage. The dark-green smaller sprouts are more desirable as they have a more delicate taste. The larger looser sprouts have a stronger flavor. Remove any yellow leaves as they indicate wilting.

In case of insect infestation, the sprouts should be soaked in a heavy, lukewarm salt-water solution for about 10 minutes. The sprouts should then be rinsed carefully in fresh water.

If Brussels sprouts are to be dried whole, make a crosscut at the base of each sprout to shorten the cooking time. Otherwise halve or cut lengthwise. Dip in bisulfite solution or another desired solution. See the section *Pretreatment of Foods.* Steam 6-10 minutes depending on the size.

Place on trays in an oven set at 140° F. for 3 hours oven door on a gas range propped open 8 inches, on an electric range half an inch. Lower the heat toward the end of drying time or keep the oven door fully open. Store in airtight containers when dry.

Drying time: 4-5 hours for halves.

To reconstitute: Cover sprouts with water, cover with a lid, and refrigerate overnight; or soak for about an hour. Bring to a boil and simmer gently until tender.

⑨ Cabbage

Cabbage comes in several varieties from the early spring conical heads to the staple winter cabbage. There is also red cabbage, deep purple in color with a tight compact head. Cabbage should be solid and heavy for its size. Avoid cabbage in which leaves have separated from the stem as they are strong in flavor and coarse in texture. The outside leaves have more vitamins so trim carefully, removing only the leaves that are yellowed, wilted, decayed, injured, or worm infested. Cabbages should be dried as soon as possible for less vitamin loss.

With a sharp knife, core then cut cabbage into shreds ¼ inch wide. Steam for 4-5 minutes or use sodium bisulfite in water to stabilize color, flavor, and odor. See the section *Pretreatment of Foods*. When steaming cabbage, place several slices of bread on top of the covered pan to help absorb some of the odor. In steaming red cabbage add a small amount of vinegar or lemon juice to help prevent color loss.

Drain well and place on trays. Dry cabbage at a low temperature of 140° F. with the oven door on a gas range propped open 8 inches, on an electric range half an inch. After several hours, either lower the heat or leave the oven door completely open. Use great care in packaging as cabbage absorbs moisture easily. Moistureproof airtight bags are a must.

Drying time: 1-3 hours for ¼ inch shreds.

To reconstitute: Cook cabbage by dropping shreds into simmering water. Cook until tender—about 15 minutes. Add salt or any seasoning desired toward the end of cooking time.

For coleslaw or fresh cabbage, place in ice water and refrigerate covered for several hours. Cabbage will not retain its former crispness.

10. Carrots

Carrots should be firm, smooth, and well shaped. The tops should be fresh and green. Very young carrots are not as good dried and require a longer drying time. Old carrots, although they contain more carotene, are sometimes woody. Carrots that are split also have this tendency and might be tasteless. Peeling is optional and great care in washing is to be exercised if they are to be scraped. Trim off any green portions. Cut into shreds, ¼ inch slices, strips, cubes, or chunks. Drop into sodium bisulfite or another desired solution to hold the color and reduce drying time. See the section _Pretreatment of Foods_. Steam blanch anywhere from 2 to 4 minutes for shreds and from 6 to 8 minutes for larger pieces. If carrots are improperly blanched they will bleed and there will be carotene loss. Place on trays in an oven set at 140° F. with the oven door on a gas range propped open 8 inches, on an electric stove half an inch. Store in airtight containers when dry.

Drying time: 6-8 hours for 1/8 inch thick slices.

"Carotene" any of several red or orange colored isomeric hydrocarbons $C_{40}H_{56}$ found in butter + in carrots + certain other vegetables + changed into vitamin A in the liver

To reconstitute: Add 1 cup dried carrots to 2 cups of water. For shredded carrots, refrigerate covered for about 15-30 minutes, then simmer for ½ hour or until tender. Add salt and seasonings the last 5 minutes of cooking time.

⑪ Cauliflower

Cauliflower is another delicious member of the cabbage family. Top-quality cauliflower is firm, creamy white in color, with leaves that are fresh and green. Stems should be peeled thinly to the base of the head and leaves removed. Soak cauliflower in heavily salted water solution to force out any insects. Cut large stalks into desired pieces and break off buds. Drop into desired solution. See the section *Pretreat-* *ment of Foods*. Use sodium bisulfite for a fresher product; it will keep cauliflower from darkening and help deter a strong odor in storage. Steam blanch for 5 minutes. Place on trays in an oven set at 140° F. The oven door on a gas range must be propped open 8 inches, on an electric range half an inch. Dry until brittle. Store in airtight containers when dry.

Drying time: 4-6 hours for serving pieces.

To reconstitute: For a crisp product to use in dips and salads, use 1 cup cauliflower to 1½ cups water. Cover and refrigerate for 30 minutes or overnight. Or soak for 30 minutes and gently simmer until tender. Season the last 5 minutes of cooking time.

(12) Celery

Celery is a biennial herb of the carrot family. The leaves and stalks are used in soups, stews, and other dishes. Celery should have firm crisp stalks and fresh bright leaves.

Wash carefully to remove dirt and any strings on coarse outer stalks. Coarser leaves may be dried and used in place of celery seeds. Place leaves on trays and dry in an open oven until crisp. Package carefully.

Cut stalks into ½ inch slices or desired pieces and steam blanch for 4-5 minutes or longer depending on the size. Pieces may be placed into desired solution. See the section *Pretreatment of Foods.* Drain. Place on trays in an oven set at 140° F. The oven door on a gas range must be propped open 8 inches, on an electric range half an inch. When dry, store in airtight containers. Do not dry other strong-flavored foods with celery unless you are making a soup mixture.

Drying time: 3-4 hours for ½ inch pieces.

To reconstitute: Add 1 part celery pieces to 3 parts water. Soak 1 hour before using.

(13) Chayotes *a tropical American perennial vine of the gourd family, grown for its edible, fleshy pear shaped, single-seeded fruit."*

Chayote is sometimes called a vegetable pear and is greatly prized for its edible tubers and the delicately flavored fruit. Chayotes are 3-4 inches long, pear-shaped, with an outer green skin, and a light inner green flesh.

Tubers are harvested after two years' growth. The young leaves of the chayote plant are sometimes used as greens. Chayotes are practically free of starch and

66

therefore are especially valued by diabetics and people)

Young and tender chayotes may be eaten without peeling. If older, they should be thinly peeled. Do not remove the seed as it is very tender when cooked.

Chayotes may be used in relishes and pickles and in the same way that you would use cucumbers. They can be added to soups and stews. If used in a salad they should be parboiled and chilled before using. Grated, they can be used as a substitute in potato pancakes. They may also replace the potato in potato salad. The cooked chayotes must be cold and chilled before adding other ingredients as they are too watery when warm.

Wash, trim, and cut into ½ inch slices. Steam for 6 minutes. Place on trays in a low oven set at 140° F. The oven door on a gas range must.be propped open 8 inches, on an electric range half an inch. Lower oven temperature at the end of drying time or leave the oven door completely open until dry. Store in airtight containers when dry.

Drying time: 4-6 hours.

To reconstitute: Add 1 cup chayotes to 2 cups water. Soak for 30 minutes and simmer until tender.

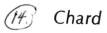 **Chard**

This plant is a form of beet grown only for its large leaves and stalks. The leaves are used as greens and have a delicate flavor rather like asparagus. The fleshy white or pale stems can be prepared like celery. Leaves should be crisp and stalks should be thick and firm. Strip stems and set aside.

15) CHARD GREENS. —
Drop chard leaves into desired solution. See the section *Pretreatment of Foods*. Steam blanch for 2-5 minutes. Longer blanching will cause the leaves to become soggy and drying time will be extended. Place on trays in an oven set at 100° F. The oven door on a gas range must be propped open 8 inches, on an electric range half an inch. Dry until brittle. Store in airtight containers.
    Drying time: 1-3 hours.
    To reconstitute greens: Add 1 cup leaves to 1 cup boiling water. Simmer gently for 2-3 minutes. Do not overcook.

16) CHARD STEMS. —
Stems should be cut into desired pieces and dipped into sodium bisulfite. Steam for a few minutes. Place on trays in an oven set at 140° F. until dry. The oven door on a gas range must be propped open 8 inches, on an electric range half an inch. Store in airtight containers when dry.
    To reconstitute stems: Add water to barely cover and soak for 30 minutes or overnight. Bring to a boil and simmer gently for 10 minutes or longer until tender.

17) Collards

Collards are a tall form of kale belonging to the cabbage family. The coarse, broad, green leaves are flavorful and prepared and eaten like greens. Sometimes cabbage seedlings (referred to as collards), are used as greens. They are grown without transplanting and harvested just before the heads are formed.

Use only young and tender leaves. Older leaves are coarse and bitter in flavor. Remove any yellowed or wilted leaves; discard any thickened stems. Leaves must be cleansed thoroughly to remove any insect spray, soil, or dirt. Change the rinse water several times. Try to remove as much water as possible, patting dry with paper towels. Care should be maintained throughout the handling process to prevent bruising and crushing. Collards must be processed quickly and within 5 hours to prevent loss of vitamins. Excess collard leaves should be held under refrigeration and dried as quickly as possible to insure vitamin content, good taste, and flavor.

Leaves should be dried whole in order to lessen vitamin loss through bleeding. Collards should be steamed and placed on trays as quickly as possible. Blanching time varies from 2 to 4 minutes or until the ribs are translucent. Overblanching will cause the soggy greens to mat and stick to the trays and drying time will be extended.

Greens may be sulfited or you may use another desired solution. See the section *Pretreatment of Foods*. Pretreatment helps the succulent collards hold their color and flavor, deters vitamin loss, and reduces drying time appreciably. The reconstituted product will have more taste and eye appeal. Drain well, spread food evenly on trays and place in oven. Avoid overcrowding.

Oven temperature should not exceed 100° F. As the product nears completion, care should be taken that the leaves do not brown. The oven door may be left completely open and contents frequently supervised to prevent browning. When dry, store in airtight containers. Do not crush leaves.

Drying time: 1-3 hours.

To reconstitute: Presoak for 30 minutes in water to cover or simply drop collards into simmering water for 3 minutes. Drain and serve.

⑱ Corn

Corn, also called sweet corn or maize, is a highly developed annual grass that has become so highly cultivated that it no longer grows in its wild state. Corn is an abundant source of starch—a very important energy food. Quality corn should have a fresh appearance, green husk, and silks that are dry and dark brown (almost black). The ear should be firm with milky kernels that are plump and juicy.

Corn may be dried several ways:

1. Husk, remove silks, and trim off any bad sections. Steam blanch from 10 to 15 minutes for the mature ears and 15 minutes longer for the very young ears. Place on trays in a 110° F. oven. The door on a gas range must be open 8 inches, on an electric range half an inch. After several hours, increase the temperature to 140° F. until dry. The kernels may then be removed with your fingers. Package carefully in airtight containers.

2. Steam blanch the corn, then cool. Using a sharp knife, cut off the kernels and the adhering particles from the cob. Spread on trays 1 inch deep. Place in a 110° F. oven. The oven door on a gas range must be open 8 inches, on an electric range half an inch. Store in airtight containers when dry.

3. *Granny's Delightful Corn Krackles:* To 4 pints raw corn, add 3 tablespoons sugar, 2 teaspoons

canning salt, and ¼ cup of sweet cream. Simmer ingredients for 15-20 minutes, stirring constantly to prevent sticking. Spread mixture over shallow pans and place in a 140° F. oven with the door on a gas stove open 8 inches, on an electric stove half an inch. The oven door should be left completely open toward the end of the drying period. Stir often and when kernels are crisp, place in paper sacks, tie securely, and hang in a dark place. The corn will rattle when completely dry. It can be stored indefinitely.

Drying time: 4-6 hours.

To reconstitute: Place in just enough water to cover, with butter and a little milk. Simmer until tender.

(19) Cucumbers

Cucumbers are grown for their fruits which are used in salads, pickles, and relishes. They are also delicious when cooked in a small amount of salted water until tender. Sliced very thin and marinated in a mildly sweetened vinegar dressing, they make a fine accompaniment to meat. They may also be dipped in beaten egg, covered with cracker crumbs or flour, and fried in deep hot fat until crisp and golden. Dried chips are delicious to eat and are especially delightful with dips.

Peel, cut cucumbers into ¼ inch slices, and place into desired solution. See the section *Pretreatment of Foods*. If you use sodium bisulfite, then keep the cucumbers in solution for 5 minutes. Drain. Place on trays in a 140° F. oven until dry. The oven door on a gas range must be open 8 inches, on an electric range

half an inch. Place in moistureproof bags when dry.
    Drying time: 4-6 hours.
    To reconstitute: Drop cucumber slices into water
and place covered in refrigerator until reconstituted.

(20) Eggplant

The eggplant, also known as Guinea squash and egg-
fruit, is a native of Southern Asia. The large, smooth,
pear-shaped fruit is used as a vegetable. The eggplants
in our market are generally the purple type. They
should be heavy and compact with a consistent color
throughout. Avoid soft, flabby, or wrinkled fruit as
they are overripe.
    Peel and slice into desired pieces no longer than ½
inch. Steam blanch for 6-8 minutes or dip into sodium
bisulfite or another desired solution and then blanch
the required time. See the section *Pretreatment of
Foods*. Place on trays in a low oven until dry and
brittle. The oven door on a gas range must be propped
open 8 inches, on an electric range half an inch.
Toward the end of drying time leave the oven door
completely open. Store in airtight containers when
dry.
    Drying time: 3-5 hours for ¼ inch pieces.
    To reconstitute: Place in water to cover. Add more
if needed. Drain well. Use in your favorite recipe.

(21) Green Peppers

Sweet peppers are green when mature, gradually
changing to a reddish gold and then to a bright red.

They are not as hot as some species of peppers and are sometimes substituted for pimiento. They vary in shape and size, from long and slender to short and thick. The latter ones are excellent for stuffed peppers if they are firm, heavy fleshed, and have an even bright color. The seed and membrane of the green pepper are much sharper than the flesh and should be removed. Sometimes the seeds are as hot and fiery as cayenne pepper. The sweet pepper is delicious when added to salads or is excellent fried. Add a few small cheese cubes to the fried peppers at the end of the cooking period. Toss carefully until cheese is melted and blended. Delicious!

Remove stems, seeds, membranes, and wash thoroughly. Slice, dice, or cut into julienne strips. They may be dipped into sodium bisulfite solution or another desired solution. See the section *Pretreatment of Foods*. Or you may steam them for 3 minutes. Place on trays in a 140° F. oven. The oven door on a gas range must be propped open 8 inches, on an electric range half an inch. Lower the oven door the last hour of drying time. Peppers dry *unevenly* (product curls and appears dry) so check them carefully before packaging. Store in airtight containers when dry.

Drying time: 2-5 hours for 3/8 inch squares.

To reconstitute: Put ¼ cup of peppers in water to cover. Chill until needed for use in fresh salads; or presoak peppers for 1 hour. They can be added dry to any recipe.

Greens

This general classification includes the leaves and stems that are sometimes steamed or boiled and served as a vegetable. The most common greens are beet tops, broccoli, thinned-out cabbage seedlings, chard, chervil, chicory, Chinese cabbage, Chinese mustard, collards, cress, dandelion, endive, escarole, kale, mustard, sorrel, spinach, and turnip sprouts. Sometimes these are known or referred to as potherbs.

Greens are a welcome sight in the spring and as the seasons change other greens vary the supply. Their young tender shoots are sometimes served raw, heightening the flavor or adding a new zest or spice to an otherwise bland food. The young leaves are delicious cooked and served with a touch of seasoning or a tad of butter. For added appeal, try a dash of paprika or simply toss croutons or chopped hard-cooked eggs over the surface for a brighter more enhancing appearance. Lemon may be added to bring out the flavor of some greens.

Use only young and tender leaves. Older leaves are coarse and bitter in flavor. Leaves must be cleansed thoroughly to remove any insect spray, soil, or dirt. Change the rinse water several times. Try to remove as much water as possible, patting dry with paper towels. Care should be maintained throughout the handling process to prevent bruising and crushing. Greens must be processed quickly to prevent loss of vitamins. Remove any yellowed or wilted leaves; discard any thickened stems. Greens should be processed within 5 hours of picking or purchase. Excess greens should be held under refrigeration and dried as quickly as possible to insure vitamin content, good taste, and flavor.

Leaves should be dried whole in order to lessen vitamin loss through bleeding. Greens should be steamed and placed on trays as quickly as possible. Blanching time varies from 2 to 4 minutes or until the ribs are translucent. Overblanching will cause the soggy greens to mat and stick to the trays and drying time will be extended.

Greens may be sulfited or you may use another desired solution. See the section *Pretreatment of* pg 10 + 11 *Foods.* Pretreatment helps the succulent greens hold their color and flavor, deters vitamin loss, and reduces drying time appreciably. The reconstituted product will have more taste and eye appeal. Drain well, spread food evenly on trays and into prescribed temperature. Place aluminum foil at oven bottom to catch drips. Avoid overcrowding.

Oven temperature should not exceed 100° F. As the product nears completion, care should be taken that the leaves do not brown. Oven door may be left completely open and contents frequently supervised to prevent browning. When dry, store in airtight containers. Do not crush leaves.

Drying time: 2-4 hours.

To reconstitute: Presoak for 30 minutes in water to cover or simply drop greens into simmering water for 3 minutes. Drain and serve.

23 *Hubbard Squash*

This is one of the largest squash of the winter varieties. It has a very dark green color that is sometimes tinged with orange and has a roughened exterior.

Remove seeds and rind. (Seeds may be cleaned and dried at a low oven temperature.) Cut into desired pieces ¼ inch or longer and steam blanch 4-6 minutes or place in desired solution and then blanch the required time. See the section *Pretreatment of Foods.* Place on trays in a 150° to 160° F. oven. The oven door on a gas range must be open 8 inches, on an electric range half an inch. Finish drying at a very low temperature or leave the oven door completely open. Store in airtight containers when dry.

Drying time: 4-5 hours for ¼ inch chunks.

To reconstitute: Add 1 cup squash to 2 cups water. Presoak 1 hour and then simmer 1 hour or until tender. It may be pureed if desired.

Jerusalem and Chinese Artichokes

The Jerusalem artichoke is a root vegetable and is seldom found in the markets as it does not keep well. However, it does grow easily in most sections of the country. Peel, trim, then cut into strips, slices, or dices. For better color, Jerusalem artichokes may be placed into sodium bisulfite solution for a second or placed into another desired solution. See the section *Pretreatment of Foods.* Blanch for 5 minutes and drain. Place on trays and in an oven set at 140° F. for several hours. The oven door on a gas range must be open 8 inches, on an electric range half an inch. Lower the temperature near the end of drying time or leave the oven door completely open. Store in airtight containers when dry.

Drying time: 4-6 hours.

The Chinese artichoke is related to the Jerusalem

artichoke so follow the instructions above.

To reconstitute: Soak in water to barely cover for several hours.

 ## Kale

A hardy biennial kind of cabbage raised primarily for its edible blue-green leaves and processed like spinach or other greens. The leaves are fine-toothed and rippled. Plants that are wilted or yellowed should be avoided although brownish-colored leaves are acceptable. Brown leaves are due to cold weather conditions during growth. Wash leaves in lukewarm water and handle carefully. Kale and other greens should not be soaked for a prolonged period of time as the precious vitamins are lost this way. Do not overcook fresh or dried kale as it will destroy the flavor. Add salt pork or bacon to the fresh or reconstituted kale for a very tasty vegetable.

Leaves should be dried whole in order to lessen vitamin loss through bleeding. Kale should be steamed and placed on trays as quickly as possible. Blanching time varies from 2 to 4 minutes or until the ribs are translucent. Overblanching will cause the soggy kale to mat and stick to the trays and drying time will be extended.

Kale may be sulfited or you may use another desired solution. See the section *Pretreatment of Foods*. Pretreatment helps the succulent greens hold their color and flavor, deters vitamin loss, and reduces drying time appreciably. The reconstituted product will have more taste and eye appeal. Drain well, spread food evenly on trays and into prescribed temperature.

Place aluminum foil at oven bottom to catch drips. Avoid overcrowding.

Oven temperature should not exceed 100° F. As the product nears completion, care should be taken that the leaves do not brown. Oven door may be left completely open and contents frequently supervised to prevent browning. When dry, store in airtight containers. Do not crush leaves.

Drying time: 2-3 hours.

To reconstitute: Presoak for 30 minutes in water to cover or simply drop kale into simmering water for 3 minutes. Drain and serve.

 ## Kohlrabi

Kohlrabi is a hardy vegetable of the cabbage family. It is sometimes called the stem turnip or turnip cabbage. It is raised for its turnip-like swollen stem but has a more delicate flavor. The colors range from pale green to purple. Young globes are steamed but older stems should be peeled and boiled. Globes larger than 3 inches become woody and strong in flavor. Peel, then cut into desired pieces and simmer in lightly salted water. The leaves of the plant may be eaten like spinach.

Wash, trim, and cut into ½ inch slices. Steam for 6 minutes. Place on trays in a low oven set at 140° F. The oven door on a gas range must be open 8 inches, on an electric range half an inch. Lower oven temperature at the end of drying time or leave the oven door completely open until dry. Store in airtight containers when dry.

Drying time: 2-3 hours.

To reconstitute: Add 1 cup kohlrabi to 2 cups water. Soak for 30 minutes and simmer until tender.

(27) Leeks

Leeks are hardy biennial herbs of the lily family, much like an onion but milder. They are grown for their enclosed stem which is used to season soups and meat dishes. The leak is the floral emblem of Wales. It is also known as the poor man's asparagus because it can be substituted in many recipes calling for asparagus. Leeks make a delicious hot or cold soup for summer meals. They are delightful cooked and served with a French dressing. For a treat, try a combination of cooked leeks and mushroom sauce. It makes a delightful gourmet vegetable.

Strip the outer leaves, leaving about 5 inches of stem and cutting off the base root. Leeks should be washed carefully as they are usually sandy. Slice stems and leaves. Place on trays in a 140° F. oven until dry. The oven door on a gas range must be propped open 8 inches, on an electric range half an inch. Store in airtight containers when dry.

Drying time: 1 or more hours.

To reconstitute: Place in water to cover and soak for 1-2 hours. Use in your favorite recipe.

(28) Lettuce

Lettuce is a hardy annual plant of the chicory family and the most important salad plant, raised for its edible leaves. Lettuce comes in several varieties including iceberg, leaf, and butterhead. Iceberg is the most popular lettuce, noted for its firmness and crispness. Leaf lettuce has either a smooth or curly leaf and is sought for its vitamins. Butterhead has a greener, smoother head and is served for its delicious flavor.

High quality lettuce is fresh looking, crisp, tender, and should be free from decay. Avoid lettuce that is going to seed as it is bitter. Lettuce that has a reddish cast to its stem will not keep long as it has been kept wet.

Lettuce will stay in its wilted state when reconstituted and can be used in wilted lettuce recipes. It is very tasty when eaten in the dry state.

Wash, trim, and chop if desired. Dip into sodium bisulfite or another desired solution. See the section *Pretreatment of Foods*. Shake off excess water and place on trays in a very low oven until dry. The oven door on a gas range must be kept open 8 inches, on an electric range half an inch. Store in airtight containers when dry.

Drying time: 1 or more hours.

To reconstitute: Cover lettuce with water and refrigerate.

Mushrooms

Food of the gods! Mushrooms have delighted man as far back as there were any written records. Hundreds of years before Christ they were exported as an item of trade. Today mushrooms are still highly esteemed. They add a festive touch to whatever food they are in. Mushrooms are the fruit of the fungus that grows below the soil. There are many varieties of edible and poisonous mushrooms and, to be safe, they should be purchased at the supermarket or from a local commercial grower. Mushrooms add elegance and fragrance to plain food. Since they contain only sixty-six calories per pound they are a boon to dieters, who can lavishly spruce up a somewhat drab meal to an epicurean delight!

Mushrooms are easily dried, and if properly packaged and stored, will keep well. They quickly deteriorate if exposed to moisture. Occasionally they should be checked for spoilage and infestation. A few peppercorns may be added to discourage insects. Mushrooms can be exposed to air for short-term storage only.

Button mushrooms are more expensive. They have a more delicate flavor than their mature brothers and are generally served whole. The caps on the more mature mushrooms have opened up and the stems are less tender. Stems are usually cut into smaller pieces and are a luxurious addition to soups, stews, scrambled eggs, rice, and fine spaghetti sauces. After drying they may be powdered and added to other foods; their pleasant taste and aroma highlighting whatever they are in. Skins, peels, trims, and tougher portions may be simmered in water to give stock a unique flavor in soups and sauces.

Prime button mushrooms should be white and without spots. They are less desirable once they begin to darken although they may be prepared and used as an addition to other foods. To dry, use only edible mushrooms that are young, fresh, firm, and clean. Do not peel the caps unless you need to remove imperfections or if the skin is coarse. Slice according to the size you desire. Buttons are usually halved or quartered while larger mushrooms are sliced in 6-10 pieces. Thinner slices require less drying time.

Cut mushrooms darken rapidly due to oxidation and this can be prevented by adding a little lemon juice, vinegar, or sodium bisulfite to the water to hold the color. Simply slice the mushrooms into the solution. Drain and place on trays in a 160° F. oven. Low heat is necessary and should be continuous, as mushrooms will mold. The oven door on a gas range must be propped open 8 inches, on an electric range half an inch. Temperatures may be lowered toward the end of drying time or the oven door may be left completely open. Do not overdry. Store in airtight containers when dry.

Drying time: 3-5 hours for slices.

To reconstitute: Dried mushrooms may simply be added to soups or sauces without soaking, as long cooking makes presoaking unnecessary. Or add 1 cup warm water to every ounce of dried mushrooms. One pound of fresh mushrooms will yield three ounces when dried. Warm water will soften tissues quickly, resulting in a softer product. Remember dried mushrooms are concentrated in flavor so do not use too lavishly in some dishes. Smaller pieces should be cooked 20 minutes or longer, larger pieces should be cooked for 30 minutes after presoaking. Presoaking liquid may be strained and added to flavor other foods.

Mustard Greens

Mustard is an annual plant of the genus *Brassica.* The

young plants' pungent leaves, tender stems, and flower heads may be used as vegetables. The leaves can be used to liven-up mildly flavored salads or combined with other green-leafed vegetables that are not as sharp and zesty.

Mustard seeds are used in pickles, relishes, and other foods. Ground seeds can be used separately or mixed together with other spices. They can be used as a flavoring or seasoning.

Leaves should be fresh, bright green, tender, and crisp. If the heads of the seeds are visible then the mustard is old and too strong.

Leaves should be dried whole in order to lessen vitamin loss through bleeding. Mustard greens should be steamed and placed on trays as quickly as possible. Blanching time varies from 2 to 4 minutes or until the ribs are translucent. Overblanching will cause the soggy greens to mat and stick to the trays and drying time will be extended.

Greens may be sulfited or you may use another desired solution. See the section *Pretreatment of Foods*. Pretreatment helps the succulent greens hold their color and flavor, deters vitamin loss, and reduces drying time appreciably. The reconstituted product will have more taste and eye appeal. Drain well, spread food evenly on trays and into prescribed temperature. Avoid overcrowding.

Oven temperature should not exceed 100° F. As the product nears completion, care should be taken that the leaves do not brown. The oven door may be left completely open and contents frequently supervised to prevent browning. When dry, store in airtight containers. Do not crush leaves.

Drying time: 2-3 hours.

To reconstitute: Presoak for 30 minutes in water to cover or simply drop greens into simmering water for 3 minutes. Drain and serve.

③ Okra (gumbo)

Okra is the king of the hibiscus or mallow family and is raised chiefly for its long ribbed pods. In the southern part of the United States it is also known as okro and more commonly, gumbo. Soups containing okra are generally known as gumbos. Okra is often served as a vegetable but more commonly used in stews, catsups, and for thickening and flavoring soups. Okra should be clean, fresh, young, and tender with pods from 2 to 4 inches long, which will snap easily when bent. Pods should be picked frequently when young as they will turn woody. Young pods can be sliced and added to soups. The formed but unripe seeds of larger pods may be shelled and used as peas. Okra is sometimes cooked whole, then chilled and served as a salad with French dressing. It can also be served in a salad after it is cooked in salt water and chilled.

To prevent discoloration, avoid cooking okra in copper, brass, iron, or tin pots. It will not affect the product but the result will lack eye appeal.

Wash, trim pods, and cut into ¼ inch widths. Steam for 5 minutes. Drain. Place on trays in a 140° F. oven. The oven door on a gas range must be open 8 inches, on an electric range half an inch. Lower the oven door near the end of drying time or leave the oven door completely open. Store in airtight containers when dry.

Drying time: 4-6 hours for ¼ inch rounds.

To reconstitute: Add 1 cup okra to 2 cups water. Use in your favorite recipe or gumbo.

(32) Onions

The onion is a hardy biennial vegetable of the lily family that is grown primarily for its firm ripe bulbs but also used for its young tender stems. From the earliest recorded time, onions have been highly esteemed. They were used to prevent thirst by travelers and soldiers on the march in the desert regions. Today they are eaten in their fresh or cooked state to add sparkle and zest in whatever they are in. Utilize these gems in whatever food you want to flavor or season.

Onions of high quality are hard, dry skinned, bright, clean, and well shaped. The colors range from white to yellow or purple. Avoid onions that are moist or rotted at the the neck of the bulb as this indicates interior decay. Misshapen onions may taste good but you may not get as much for your money.

The onion odor is so cohesive that you may wish to have a small chopping block reserved for it and its cousin, garlic. To remove the odor, sprinkle the board with salt and rinse. The volatile oils that cause tearing are easily eliminated by either blanching for two minutes (the final product will not be as good) or holding the onions under running water while preparing them. There are several methods used to remove their clinging odor from the hands. Either rub a generous amount of lemon juice into the hands or make a paste of baking soda and water. Salt or even dry mustard may be rubbed into the hands and then rinsed.

Onions are one of the simplest vegetables to prepare. In their cooked state they do lose some of their potency but retain their fiery bite when served crisp in their dried state and also when used in a dip.

Blanching is unnecessary. Onions need only to have the skins and crown removed. Slice or dice and place on trays in a 140° F. oven. The oven door on a gas range must be open 8 inches, on an electric range half an inch. Onions brown easily so leave the oven door completely open at the end of the drying period. Store in airtight containers when dry.

The simplest way to process onions is to slice them into rings and place them on trays. As they dry, the heat gradually shrinks the membranes and one has perfect rings that need only to be turned to dry evenly. Removal is simple. This method does save space.

Drying time: 3-6 hours for ¼ inch slices.

To reconstitute: Merely cover flakes, slices, or dices with water and use after soaking 1-2 hours depending on the size and whether they are blanched or unblanched. Reserve any leftover liquid as this can be added to other ingredients for flavoring. In recipes that contain a small amount of liquid, presoak only for a few minutes in a small amount of water before adding to food.

⟨33⟩ Parsnips

Parsnips are biennial herbs, grown for their thick, long white root. Parsnips are prepared like carrots and similar in shape but do not have their taste or color. They are winter vegetables that need to be nipped by the frost to bring out their distinctive flavor. Parsnips are at their best when their roots are small to medium size, firm, and well shaped. Larger plants have woody cores that must be removed. Avoid roots that are shriveled and soft.

Parsnips should be steamed to bring out their sweet, nutty flavor. Cut lengthwise and remove the core; it is tasteless and does not contribute to the flavor. Cut into desired pieces. Drop into sodium bisulfite or another solution. See the section *Pretreatment of Foods*. Steam blanch anywhere from 2 to 4 minutes for shreds and from 6 to 8 minutes for larger pieces. Place on trays in an oven set at 140° F. with the oven door on a gas range open 8 inches, on an electric range half an inch. Shredded parsnip will dry in several hours depending on amount dried. Store in airtight containers when dry.

Drying time: 4-5 hours for ½ inch slices.

To reconstitute: Add 1 cup parsnips to 2 cups water. Soak for 15-30 minutes, then cook for 5-10 minutes or until tender.

34 / Peas

Peas are an annual, herbaceous, climbing vine, grown mainly for their edible seeds and pods belonging to the genus *Pisum*. Peas have a romantic history, that according to some authorities, goes back to the Stone Age. During the Middle Ages these legumes were grown as a protection against famine and were added to the diet of the armies of that time. Their oldest use was in the form of a dried seed and it wasn't until the Middle Ages that people began using the fresh pods and peas.

Peas make a good rotation crop for sweet corn. For

planting peas, fertilize the ground the preceding fall.

The choicest peas are bright green in color, fresh, fully developed, young, tender, and sweet. Peas should be picked frequently as they lose their sweetness when they reach maturity. Steaming is preferred but if cooked in a little water, sugar may be added. They are delicious when served with a sprinkle of mint.

Shell. Steam 3-5 minutes. Peas must be rinsed in cold water after steaming. Place on trays in a 150° F. oven. The oven door on a gas range must be open 8 inches, on an electric range half an inch. Lower temperature to 140° F. or keep the oven door completely open during the end of drying time. Store in airtight containers when dry.

Do not discard washed fresh pods as they can be used for coloring soups to replace commercial coloring. Lay pods on cookie sheets and place in a 300° F. oven until dry and brown. Turn to brown evenly. Do not allow to burn. Keep in an airtight container. They will last indefinitely. Add 4 pods to bouillon or consomme for coloring pale stock.

Drying time: 3-5 hours.

To reconstitute: Add 2/3 cup peas to 2 cups water. Presoak for 2 hours. Cook and add salt during the last few minutes of cooking time. Drain and season to taste. Do not overcook as long blanching causes skins to rupture.

35. Pimientos

The term pimiento is sometimes applied to various forms of capsicum. At times, green pepper is allowed to mature into a deep red and is prepared and used as pimiento. Remove the seeds and membrane and wash carefully. Cut into desired pieces. They may be placed on trays and in a low oven or steam blanched for 3 minutes. They can also be sulfited. See the section *Pretreatment of Foods.* The oven door may be left open during the latter part of drying time. Peppers dry *unevenly* so check carefully before packaging. When dry, pepper should be bright red in color with no scorching or browning. Pimientos may be ground for paprika if desired.

Drying time: 2-5 hours for 3/8 inch squares.

To reconstitute: Add 2 tablespoons pimiento to ½ cup water. (Double or triple proportions.) Soak for 1 hour. For salads: Cover with water and chill in refrigerator until crisp and juicy.

(36) Potatoes

Potatoes are a tropical and subtropical perennial herb. They are closely related to the eggplant, pepper (capsicum), and tomato. The potato is one of the most popular vegetables because of the many ways it can be prepared and blended with other foods. Because of its starchy composition it is one of the first foods dieters shy away from, and yet rice and peas and other legumes are just as starchy and as fattening. The starch of the potato is more digestible than many other starches.

In many places this staple, like bread and rice, is considered the staff of life. It can make a humdrum meal a gourmet's delight. The potato takes on whatever food it is with and blends itself in perfect harmony with its flavor, yet it can stand alone in all its glory. Have it fried, boiled, mashed, or baked—it still comes out a winner.

Choose potatoes that are firm, well aged, smooth skinned, reasonably clean, and white and mealy when cooked. Avoid potatoes that have a touch of green, as this condition is caused by prolonged exposure to light, either in its growing stage or through long storage. This area is very bitter. It is the same substance that is found in sprouting potatoes. Potatoes must be kept in the dark to keep from sprouting. Potatoes that are discolored or wilted should be avoided as they either have been harvested too early or were improperly stored.

Potatoes are simple to prepare for drying, though care should be taken to keep them from turning black (oxidation). Raw potatoes may be placed into a sodium bisulfite solution for 10 minutes and onto the trays to dry. Raw potatoes prepared this way take longer to dry, reconstitute, and then longer to cook.

Steam blanching is the better method for preparing potatoes for drying. They can be placed into a desired solution to prevent darkening and then steam blanched. See the section *Pretreatment of Foods*.

Riced potatoes: Boil potatoes in their jackets until done. Cool, then peel and place into a ricer or strainer. Do not attempt to process them unless they are cold, as they will form a gummy mass.

Diced, julienne, or sliced potatoes should be steam blanched for 3-6 minutes or longer until they are firm but tender. Blanch over medium-high heat. Smaller pieces should be blanched according to their size. A good rule of thumb is to check for the individual cut and type of potato. For most dices, 3-6 minutes should be adequate. Do not overcook as the product will disintegrate during presoaking treatment or in cooking. If the product is too raw, it will take longer to cook and the result will be a less flavorful product. Potatoes must be rinsed in cold water after steaming. Place on trays and into the oven at 140° F. Temperature should not exceed 150° F. The oven door on a gas range must be open 8 inches, on an electric range half an inch. Leave the oven door completely open toward the end of drying time. When dry, store in airtight containers.

Drying time: 4-6 hours for dices.

To reconstitute: If the potatoes have been dried without blanching, then presoak for at least 30-60 minutes. Add 1 cup potatoes to 2 cups water.

Preblanched dices, julienne strips, and slices should be presoaked about 15-30 minutes. Placing in ice water will result in a much firmer product. Drain off any surplus liquid if you plan to fry them.

Riced potatoes should be merely dropped into boiling water. Use from 4½ to 5 cups water to 1 cup riced potatoes. Stir constantly for 15 minutes.

(37) Pumpkin

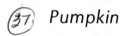

The pumpkin is related to the cucumber and belongs to the gourd family. Fruit should be left on the vine to cure for 10 days before storing. Skin should be hard and difficult to puncture. The pulp should be dark yellow, fine grained, and heavy. Smaller pumpkins are preferred—not more than 12 inches in diameter

Remove seeds and rind. (Seeds may be cleaned and dried at a low oven temperature.) Cut into desired pieces ¼ inch or longer and steam blanch 4-6 minutes or place into desired solution and then blanch the required time. See the section *Pretreatment of Foods*. Place on trays in a 150° to 160° F. oven. The oven door on a gas range must be open 8 inches, on an electric range half an inch. Finish drying at a very low temperature or leave the oven door completely open. Store in airtight containers when dry.

Drying time: 3-4 hours for ¼ inch lengths.

To reconstitute: Add 1 cup dried pumpkin to 2 cups water. Presoak 1 hour and simmer 1 hour or until tender. Pumpkins may be pureed if desired. Pumpkin can be added to chopped meats, combined with pureed carrots, used in soups, and added to breads and cookies.

 ## Rutabagas

Rutabagas are a hardy biennial herb cultivated for their tubers. The rutabaga is also known as yellow turnip and is sometimes called a Swedish or Russian turnip. Its very large, tough, thick root is the only edible section. Rutabagas are prepared like turnips. Wash, peel, and cut into slices, dices, or julienne strips. See the section *Pretreatment of Foods* for desired solution. Blanch for 5-10 minutes, depending on the size. Place in a 140° F. oven. The oven door on a gas range must be open 8 inches, on an electric range half an inch. Rutabagas sometimes darken during the latter part of drying. This does not affect the product. When dry, store in airtight containers in a cool dry place.

Drying time: 4-6 hours.

To reconstitute: Add 1 cup rutabagas to 2 cups water. Presoak for 1 hour, then simmer for 5 minutes or until tender. Do not overcook. Drain and season to taste.

39 Spinach

Spinach is a short season potherb, the greens of which can be used either raw or cooked. Spinach is healthful and delicious but can become most unappetizing if not properly prepared.

The stocky plant should have dark green crisp leaves. Do not use yellowed, wilted leaves, or spinach that has bolted to seed, as it is overgrown and tough. Young spinach can be used fresh in salads and is delicious mixed with other greens. Use with a favorite dressing for a delightful change; French dressing is excellent.

Spinach should be washed several times in lukewarm water to remove sand and any particles clinging to its leaves. The large tough stem should be removed. Leaves can be treated in one of the solutions listed in the section *Pretreatment of Foods.* Spinach can be chopped or left whole. Blanch 2-4 minutes and rinse in cold water. Place on trays in an oven set at 140° F. The oven door on a gas range must be open 8 inches, on an electric range half an inch. Leaves dry quickly depending of course on the amount and the moistness of the product. Store in airtight containers when dry. Avoid crushing.

Drying time: 1 or more hours.

To reconstitute: Barely cover with water and soak overnight. Or soak for 30 minutes, bring to a boil, and simmer 10 minutes or longer until tender. A dash of nutmeg or mace adds a distinctive flavor. It can also be seasoned with a dash of vinegar or a touch of lemon. Do not use much salt to season as dried spinach tastes salty after reconstitution.

40) Summer Squash

Squash and gourds are also included in the cucurbits and are either under the summer or winter classification. Pattyman, or yellow squash (generally crookneck), and zucchini are summer squash, best picked and marketed in their young state. They are at their prime and delicately flavored when the rind is tender and the seeds small. They can then be eaten whole.

Mature squash have hard seeds and flesh that is apt to be tough and stringy. Squash should be steamed to retain flavor. Add a small amount of water to allow the squash to steam in its own juice. The flesh should be firm and almost translucent. Season lightly with salt, pepper, and butter. Summer squash may also be baked, fried, or served in small chunks in salads. Peel and seed older squash before cooking. All squash should be firm and heavy with a fresh, bright appearance and free from bruises and decay. Squash should be handled carefully and held under refrigeration until you are ready to dry them.

Peel and slice into desired pieces no longer than ½ inch. Steam blanch for 6-8 minutes or dip into sodium bisulfite or desired solution and then blanch the required time. See the section *Pretreatment of Foods*. Place on trays in a low oven until dry and brittle. The oven door on a gas range must be open 8 inches, on an electric range half an inch. Toward the end of drying time leave the door completely open. Store in air-

94

tight containers when dry.

Drying time: 4-6 hours.

To reconstitute: Place in water to cover for 1 hour. Add more if needed. Drain well. Use in your favorite recipe.

(4/) Sweet Potatoes

The sweet potato is a very important tropical perennial trailing herb grown for its tubers. It is a very popular winter vegetable. There are two types of sweet potatoes. One is a pale yellow color with a slight yellowish brown skin with flesh that is dry and mealy when cooked. The other has a deep orange color with a red skin and its cooked pulp is moist and sweet. Good quality tubers are smooth, well developed, and free from blemishes and bruises. Sweet potatoes with soft areas should be avoided as the soft area sometimes taints the flesh of the remainder of the root.

Sweet potatoes are delicious baked, and for a change of pace, try adding a dash of Tabasco sauce to the pulp or scoop out the inner portion leaving a good sturdy shell. Mash the pulp with a tablespoon of peanut butter and replace in shell. Dot with butter and return to oven until heated through. Delicious!

Peel and cut tubers into desired pieces. Sweet potatoes will darken unless dropped into sodium bisulfite or another desired solution. See the section *Pretreatment of Foods.* Soaking in a solution will also rinse off excess starch. Steam blanch for 6 minutes or until pieces are translucent. Cooked sweet potatoes may be riced directly onto trays. Do not attempt to handle to prevent mashing and matting. Place on trays in a 160° F. oven. The oven door on a gas range must be open 8 inches, on an electric range half an inch. The oven door may be left completely open during the latter part of drying.

Drying time: 4-6 hours.

To reconstitute: Presoak for 1 hour in water to cover; then simmer for 30 minutes. Overcooking will cause the pieces to deteriorate, while undercooking will give the product a raw taste with a loss of flavor.

Swiss Chard

See *Chard*.

Tomatoes

Tomatoes are a tropical herb related to the eggplant, potato, and pepper (capsicum)—an excellent source of vitamin A and C. Tomatoes are generally thought to come from South America, but some authorities claim they came from Africa, while others from China or India. Tomatoes were primarily used as a decorative plant and considered poisonous. They are also called paradise apple or love apple.

Vine-ripened tomatoes are outstanding in flavor. Green tomatoes may be sun-ripened but do not have the taste and appearance that their vine-ripened brothers have. Ripe fruit should be bright red, fresh in appearance, firm, and heavy. They may be sulfited to preserve color and vitamins. They may be dried either with or without the peel and seeds. If not properly packaged, the oil in the tomato seeds will turn rancid.

Cherry tomatoes need only to be washed and

halved. Place on trays in a 140° F. oven. The oven door on a gas range must be propped open 8 inches, on an electric range half an inch. Paper toweling may be placed under trays and aluminum foil at oven bottom to catch drips. Watch paper toweling carefully to be sure it doesn't burn. Paper toweling may be removed when food firms up. When dry, package carefully in airtight containers; or place in a blender for tomato powder and then package. Store in a cool dry place.

Larger tomatoes may be peeled, seeded, and cut in desired chunks or slices. Follow the same instructions for cherry tomatoes given above.

Drying time: 6-8 hours for ¾ inch slices.

Dried or powdered tomatoes may be used in stews, soups, spaghetti, sauces, or any one of your favorite recipes. Dried tomatoes are more concentrated in flavor and should be used sparingly as a seasoning. A tiny bit of sugar may be added to bring out the flavor and to cut down on the acid. Tomatoes do not reconstitute as well as other vegetables and should be soaked for at least 8 hours before using. The reconstituted product will be soft.

To reconstitute:

Tomato Soup: Add 1 cup tomato powder to 4 cups water and ¼ cup dry milk.

Tomato Paste: Add 1 cup tomato powder to 2 cups water.

Tomato Sauce: Add 1 cup tomato powder to 3 cups water—season if desired.

Tomato Juice: Add 1 cup tomato powder to 4 or 5 cups water. Season with salt, pepper, lemon, or lime.

Place any of the above into the blender until smooth. Since tomatoes are composed mainly of liquid, the yield is small.

(44) Turnips

The turnip is a semihardy biennial herb grown as an annual and raised mainly for its enlarged roots which serve as vegetables. A member of the mustard family, it is a neglected vegetable. Like many other root vegetables, it must be peeled before cooking. Turnips are at their best in the immature state. The leaves should be clean, crisp, and tender while the small almost round spheres should have only a few fibrous roots at the base. Large coarse leaves are likely to be tough and stringy while the large overgrown turnips are apt to be woody and strong-flavored. Turnips may be eaten raw or cooked gently, only long enough to make them tender. They may be glazed, french fried, or baked. They enhance the flavor of soups and stews, at the same time improving their own flavor. Mashed turnip touched with grated cheese and browned is another delicious way of serving this hardy vegetable.

To dry turnip roots, cut into shreds, ¼ inch slices, strips, cubes, or chunks. Drop into sodium bisulfite solution or another desired solution. See the section *Pretreatment of Foods*. Steam blanch anywhere from 2 to 4 minutes for shreds and from 6 to 8 minutes for larger pieces. If turnips are improperly blanched there will be a loss of vitamins. Place on trays in an oven set at 140° F. with the oven door on a gas range open 8 inches, on an electric range half an inch. Store in air-tight containers when dry.

Drying time: 4-8 hours.

To reconstitute: Add 1 cup dried turnips to 2 cups water. For fresh or shredded turnips, refrigerate covered for about 15-30 minutes, then simmer for ½ hour or until tender. Add salt and seasonings the last

5 minutes of cooking time.

See also *Greens*. Turnip greens are usually called sprouts as only the green tops of young turnips are used.

Winter Squash

There are three main varieties of winter squash: acorn, marrow, and hubbard. Unlike summer squash, winter squash is allowed to ripen on the vine until threatened by the frost. It can be stored for long periods of time. The rind and seeds are hard but the flesh is tender. The stringy portion is minimal and can be easily removed along with the seeds. Seeds may also be washed and dried. Winter squash is a good substitute for pumpkin.

Acorn squash is small, dark green, and shaped like an acorn. It is too stringy when it is streaked with orange. Acorn squash can be halved and placed in a shallow pan with a minute amount of water and baked until tender. Or you can also place them face down and bake for 15 minutes, then turn squash face side up and stuff with sausage, ground pork, or dressing. Bake until tender.

Marrow is an oblong, thick-fleshed squash of excellent quality.

Hubbard squash is the largest of all, with a rough, dark-green color that is sometimes tinged with orange.

Remove seeds and rind. (Seeds may be cleaned and dried at a low temperature.) Cut into desired pieces ¼ inch or longer and steam blanch 4-6 minutes or place in desired solution and then blanch the required time. See the section *Pretreatment of Foods*. Place on trays in a 150° to 160° F. oven. The oven door on a gas range must be propped open 8 inches, on an electric range half an inch. Finish drying at a very low temperature or leave the oven door completely open. Store in airtight containers when dry.

To reconstitute: Add 1 cup squash to 2 cups water. Presoak 1 hour and then simmer 1 hour or until tender. It may be pureed if desired.

 ## Zucchini Squash

Zucchini is a member of the cucurbit family, which includes pumpkins, gourds, and squashes. While there are many different varieties of squashes, they are roughly divided into two types—the summer and winter varieties. Zucchini is sometimes known as Italian squash or vegetable marrow. It is marketed when young and tender and may be eaten in its entirety. It is for this reason zucchini should be fresh, clean, free from defects, and tender enough that a nail pierces the skin easily. They should be handled carefully to prevent bruising as they will decay quickly. Overripe squash has a tough rind, is seedy and stringy.

Zucchini should be stored in the refrigerator until used and then covered and steamed to preserve its delicate flavor. It may be cut into smaller pieces and added raw to salads or cut into finger lengths to use in dips. The latter is especially delicious when served with sour cream mixed with a Roquefort cheese dressing.

Peel and slice into desired pieces no longer than ½ inch. Steam blanch for 6-8 minutes or dip into sodium bisulfite or desired solution. See the section *Pretreat-ment of Foods*. Place on trays in a low oven until dry and brittle. The oven door on a gas range must be open 8 inches, on an electric range half an inch. Toward the end of drying time leave the oven door completely open. Store in airtight containers when dry.

Drying time: 3-4 hours.

To reconstitute: Place in water to cover. Add more if desired. Drain well. Use in your favorite recipe.

ges fm pg 102 to 106 = 5 pgs   chpt. 7.

Drying Fish

Dried fish is not only appealing for its concentrated food value but is tasty, nutritious, and stimulates the appetite.

Fish deteriorate and decompose very rapidly due to outer heat or through the enzymes within the flesh. If the situation is allowed to continue this will result in flavor loss and eventual spoilage. Fishermen hold their catch under running water to insure freshness. At times, the fish are cut, cleaned, and held in water until proper refrigeration is available. Icing retards spoilage for short intervals only. Fish should always be firm, with very little slime. There should be a fresh odor clinging to its exterior. Your nose should guide you to this freshness! Under refrigeration, fish should be carefully wrapped to prevent loss of freshness and flavor. Other refrigerated foods must also be protected from absorbing fish odors and flavors.

All fish should be carefully washed in clean water to remove outer particles of blood and slime. The following brine solution is adequate for 6 pounds of fish: Add ½ cup of salt to 8 cups of water and soak for one hour. Rewash; then resoak in a stronger brine solution, using 4 cups of salt to 6 cups of water or 2/3 cup salt for every cup of water used. This procedure will act as a preservative to draw out the moisture and concentrate the amino acids. Other spices and flavorings may be spread on the fish after the basic preliminaries have been completed. Spices add a touch of zest to the flavor of the fish and at the same time contribute to their preservation.

Place on trays in a 170° F. oven for 24 hours or dry to at least 15% of its original moisture content. If drying time must be interrupted, freeze and continue processing after fish defrosts. The oven door on a gas range must be open 8 inches, on an electric range half an inch. Be especially careful to complete the drying schedule or spoilage may occur; the inner bacteria will activate and multiply quickly if any moisture is retained. When dry, package in moistureproof containers and store carefully. Check periodically for any change in color or sign of moisture. Fish can be frozen for long periods if desired.

Remember to use lemon juice in cleaning and deodorizing utensils, sink, etc.

Dry Cure For Fish

¼ cup salt
¼ cup brown sugar, honey, or 2 tablespoons molasses
1 teaspoon liquid garlic
1 teaspoon liquid onion
Dash of paprika or cayenne if desired
6 pounds of fish

Other spices may be added to suit the individual taste. Sprinkle lightly over the fish. Place on trays in a

170° F. oven for 24 hours or dry to at least 15% of its original moisture content. The oven door on a gas range must be open 8 inches, on an electric range half an inch. Be especially careful to complete the drying schedule or spoilage may occur. Package in airtight containers and store carefully. Check periodically for any change in color or moisture content.

The following medium cure is also good for shellfish and salmon. Mixture should stand 24 hours to blend flavors:

½ pound salt
¼ pound sugar
¼ teaspoon onion powder
1 heaping teaspoon white pepper
¾ teaspoon garlic powder
6 pounds of fish

Sprinkle lightly over fish. Place on trays in a 170° F. oven for 24 hours or dry to at least 15% of its original moisture content. The oven door on a gas range must be open at least 8 inches, on an electric range half an inch. Be especially careful to complete the drying schedule or spoilage may occur. Package in airtight containers and store carefully. Check periodically for any change in color or moisture content.

Salmon

This large game fish is in season from May to September. It has an excellent flesh with an orange-pink color when cooked. It is extremely delicious and nutritious and is eaten fresh, canned, salted, smoked, or dried. It can be baked, boiled, broiled, or sautéed.

If salmon is boiled it should be allowed to cook in

its own broth or liquid to preserve its quality and flavor. Skin is easily removed when salmon is lukewarm. It is then garnished and usually served with a sauce. A green or vegetable salad makes a nice accompaniment for salmon. Capers, anchovies, and hollandaise sauce also highlight this popular and tasty fish.

To dry: Cut salmon into 3/4 inch strips and soak for 1 hour in a brine solution consisting of ¼ cup salt to 1 quart of water. Drain and spread with one of the cures listed above. Place on trays in a 170° F. oven for 24 hours or dry to at least 15% of its original moisture content. The oven door on a gas range must be open 8 inches, on an electric range half an inch. Be especially careful to complete the drying schedule or spoilage may occur. Package carefully in airtight containers and store carefully. Check periodically for any change in color or moisture content.

Salmon bits are delightful when added to canapés, served on crackers, or eaten as a snack. Try salmon chowder for a delicate, delightful treat. Serve with tiny crackers, diced potatoes, bread cubes, or croutons.

Shrimp

Wonderful crustacean! Shrimp are especially high in minerals, which are body building elements. Shrimp are easily prepared and there are several ways to prepare them.

1. Wash shrimp and place into a salt brine consisting of 1 cup salt to 8 cups water. Let shrimp simmer for 10 minutes or until the shells separate from the meat.

Remove the shells and black intestine running down the center of the back. Place on trays in a 170° F. open oven until dry. Package carefully in moistureproof bags or containers. They may be refrigerated or placed in the freezer.

2. After preparation use any desired beef jerky recipe. See the section *Drying Meats* for desired cure. Place on trays in a 170° F. open oven until dry. Package carefully in moistureproof bags or containers. They may be refrigerated or placed in the freezer.

3. After preparation use any dry cure for fish. See the section *Drying Fish*. Place on trays in a 170° F. open oven until dry. Package carefully in moistureproof bags or containers. They may be refrigerated or placed in the freezer.

4. After preparation, place shrimp into 1 quart water. Add 2 cups minced celery, 1 tablespoon whole black pepper, 1 tablespoon salt and ¼ cup chopped onion. Barely simmer for 15-20 minutes. Drain; then dry. Place on trays in a 170° F. open oven until dry. Package carefully in moistureproof bags or containers. They may be refrigerated or placed in the freezer.

Drying time: 2-5 hours.

chpt. 8.

gre fm pg 107 to 112 = 6 pp,

Drying Meats

Jerky can be made from almost any type of meat although flank and round steak are the most commonly used cuts. The cheaper cuts, such as shank or chuck, can also be processed; with a little care they will result in an excellent jerky. Pork and wild game must be steamed to remove redness, a protection against the parasitic trichinae commonly known as trichinosis. Leftover cooked meats, and this includes poultry, may also be dried for later use. They may be seasoned if desired or these dried tasty bits may be chopped or placed into the blender and powdered for adding to stews, soups, stocks, or pemmican.

Meats that are merely seasoned and placed on trays to dry will take longer to dry and will not reconstitute as quickly as the precooked meats. Meat slices may be steamed for 30 minutes, stirring to insure even steaming. (This is imperative with pork or wild game.) Steaming will stop the enzyme action, prevent growth of bacteria, concentrate the proteins, and remove part of the moisture.

Meats should be sliced with the grain and cut into ¼ or ½ inch slices. All visible fat should be removed.

Combine spices and marinate overnight in one of the marinades given below. Drain and place on trays in a 120° F. oven for 9 or 10 hours. The oven door on a gas range must be open 8 inches, on an electric range half an inch. Aluminum foil should be placed at the oven bottom to catch drips. When dry, strips will be black and will crack but not break when bent. Package carefully in airtight containers.

Meat can also be brined or dry cured. Unless brined, do not salt the meat as this will toughen it. Instead, use sugar, herbs, and spices to enhance the flavor and tenderness of the finished product. Monosodium glutamate and meat tenderizers may also be used to bring out the flavor and soften the tissues.

If meats are brined for long periods, hold under refrigeration. Meat must be immersed in liquid and held under with a saucer or plate to keep the meat completely submerged. All equipment must be sterilized to prevent bacteria from forming.

There are numerous ways of brining meats. Try a favorite sweet pickle recipe for a slightly different but tasty jerky. Soak slices for an hour and apply any desired dry seasoning. The instructions for drying jerkys are given after the last recipe in this chapter.

Or meat slices may also be soaked in a brine solution consisting of 1 pound salt to 2 quarts water. This brine should be strong enough to float an egg. Hold meat in brine for 2 days. When brining is completed, rinse, pat dry with paper towels, and add desired seasonings to both sides of the meat. Avoid using any salt. The meat has absorbed enough during the brining process—even powdered garlic and powdered onion should be substituted for garlic salt and onion salt. All of the following jerkys are dried the same way. The instructions are given after the last recipe in this chapter.

To Dry Cure Meats

6 pounds of meat
½ pound of canning salt
¼ cup sugar
1 teaspoon garlic powder
1 teaspoon onion powder
1½ tablespoons pickling spices

Mix ingredients and allow flavors to blend for 24 hours. Apply ¾ of the mixture to both sides of the meat, rubbing well into the surface. Refrigerate at 40° F. for 1 or 2 days. Work in the remainder of the mixture and refrigerate another 24 hours. Meat may be held for longer periods of refrigeration if desired. Rinse lightly, pat dry, and follow the instructions for drying at the end of this chapter.

Zesty Jerky

6 pounds flank steak or desired cut
2 tablespoons salt
2 teaspoons garlic powder
2 tablespoons onion powder
Water to cover
Optional:
Monosodium glutamate (to bring out the flavor)
Tabasco sauce (to taste)
Soy sauce (to taste)
2 teaspoons paprika

Mix and pour over sliced meat. Let stand 2 hours in the refrigerator. Follow the instructions for drying given at the end of this chapter.

Soy Marinade

6 pounds flank steak or desired cut
2 teaspoons salt
½ teaspoon ground pepper
½ teaspoon garlic powder
½ teaspoon ground ginger
2 tablespoons sugar
1 cup soy sauce
Mix and pour over sliced meat. Let stand 24 hours in refrigerator, then follow the instructions given for drying at the end of this chapter.

Spiced Jerky

6 pounds flank steak or desired cut
2 tablespoons salt
2 teaspoons ground pepper
1 teaspoon basil
½ teaspoon marjoram
½ teaspoon oregano
½ teaspoon thyme
Combine and spread over sliced meat. Refrigerate for several hours. Follow the instructions given for drying at the end of this chapter.

Italian Marinade

6 pounds flank steak or desired cut
2 teaspoons pepper
1 teaspoon garlic powder
½ teaspoon onion powder
1 teaspoon basil
1 teaspoon marjoram
½ teaspoon oregano
½ teaspoon thyme
4 tablespoons soy sauce
2 cups water

Combine and pour over sliced meat. Refrigerate for 2 hours. Follow the instructions given for drying at the end of this chapter.

Pemmican

The following recipe makes an excellent snack for camping, hiking, or boating.
8 ounces dried beef
8 ounces of any of the chopped dried fruit (apricots, currants, peaches, pears, and raisins)
8 ounces finely chopped nuts of your choice
4 tablespoons peanut butter
1 tablespoon honey
½ teaspoon cayenne pepper

Pulverize meat or grind in blender. Sprinkle and mix in cayenne pepper. Add the chopped dried fruit and mix. Heat the honey and peanut butter over *very low heat*. Add to mixture, blending carefully. Be sure contents are thoroughly mixed. Form into oblong roll. It may be dipped in warm paraffin if desired. Keep in a cool dry place. It should last indefinitely. Remove paraffin before consuming.

For a basic but delicious stew, add flour and water to pemmican and slowly simmer until hot, bubbly, and tender.

DRYING.—

Drain and place on trays in a 120° F. oven for 9 or 10 hours. The oven door on a gas range must be open 8 inches, on an electric range half an inch. Aluminum foil may be placed at the oven bottom to catch drips. When dry, strips will be black and will crack but not break when bent. Package carefully in airtight containers.

With proper care beef jerky, cured meats, and pemmican should hold indefinitely. If moisture or other conditions are introduced the product will not hold. Check periodically for any problem. If in doubt, discard food.

copyo 7/1976

gres fm fg. 113 to 142 =
(30 pgs) *chpt.* **9.**

Herbs and Spices

gres fm fg. 113 to 142 ...

Centuries ago, herbs and spices were so highly esteemed that the search and exploration for them resulted in the discovery of the New World. In addition to their primary use in preserving, seasoning, and flavoring foods, they were also valued for their medicinal properties and use as a stimulant. Teas were brewed from certain leaves and some of the more fragrant ones were used in perfumes and sachets. Others made excellent garden repellents.

Some herbs and spices were credited with strange powers. The ancient Romans and Greeks wore parsley to protect them against wine fumes and to deter intoxication. Thyme was known as a symbol of courage, basil was used to attract sympathy, chervil to stop hiccups, and dill was said to keep one from falling asleep. Today, herbs and spices are used mainly to add color to foods and make them more tasty and appealing. Some herbs are still used as teas and some medicinally, but to a much lesser degree.

Spices come from various sections of certain trees, the parts that are the richest in flavor. This includes the bud, fruit, seed, root, or bark. They are used to sharpen the appetite and to aid digestion. Spices are very pungent and should be used sparingly to lightly flavor food, never to overpower it. If properly packaged and the oils protected, the spices should last for at least one year. Both herbs and spices should be quickly processed to preserve fragrance and color.

Leaves, before they flower, will have a higher oil content and more flavor and aroma. Some of the common leaves in use are bay leaves, mint, basil, and parsley. They should be picked on a hot, sunny day as soon as the dew is gone. Care should be taken in handling and washing the leaves as they bruise easily, causing a loss of oil and discolored leaves when dry. If leaves are dusty, run them under cold water and shake off the excess moisture. Remove any blemished or bruised leaves. Place on trays in a 100° F. oven with the oven door completely open. Be sure the air is circulating freely. Leaves should take just a few hours to dry. Check every 15 minutes. Rotate trays frequently. Turn leaves for faster drying. Leaves may be crushed before packaging but will retain their oils better if packaged whole. Package in airtight moistureproof containers to keep flavor freshness. Leaves should be checked for moisture or mold after one week of storage. If mold has developed, leaves are not fit to eat or use in recipes.

Flower Heads can be dried whole, and if only the petals are used, remove them from the calyx and follow the outline under drying leaves. Sweet marjoram leaves, stems, and flowers are all used. Others include saffron, borage, clary, and hyssop.

Seed Heads should be cut as soon as the seeds are dry. This condition is noticeable in some plants when the seeds or pods change color. This usually occurs

114

just before they scatter. Commonly used seeds include poppy seeds, dill seeds, and caraway seeds. If you are unsure about the dry state of mature seeds, place the seeds in an open oven for a few hours. Package whole and grind the seeds as needed. They deteriorate very rapidly when the outer casing is cracked. Properly packaged, the seeds will keep for at least one year.

Roots should be harvested in the fall or winter. There is an abundance of concentrated food in roots. Commonly used roots include turmeric, horseradish, and shallots. Cut only a few shoots from each plant. Roots should be washed and trimmed to remove any dirt or slight blemishes. Slice and cut lengthwise and dry in a low, open oven. Drying will take place after several hours. Package carefully. Roots should be checked and reprocessed if there is any moisture in the container.

Today, there is a revived interest in using herbs as pesticides and fungicides. This is a natural way of controlling pests, without contaminating the soil and atmosphere. A few of these preventative herbs are presented here for use in protecting against infestation of garden predators.

Tansy, if grown near the kitchen, will repel ants. Ground anise and coriander seeds contain an oil that makes a very effective spray against aphids and spider mites. Garlic, when planted under the trees, will also deter aphids and peach borers from invading. Garlic, when used as a spray, will destroy bacteria and fungus that damages fruits and vegetables. Sage and rosemary, if planted near a cabbage patch, will successfully discourage the white butterfly. There are many, many others. Your public library is a good place to check for more information on herbs. Also check your local newspapers and magazines for other helpful, natural pesticides.

Anise

Anise is an annual herb that is grown primarily for its seeds, which have a mild licorice taste. The seeds are used for flavoring candy, cookies, pastry, breads, and some cheeses. The leaves can be used as a vegetable, garnish, or to season other foods. Anise is sometimes served in a very pleasing and soothing summertime drink. A teaspoon of fresh anise, mixed with a little honey and added to warm milk, will help induce sleep. It is also used in medicine as a carminative.

Strip washed leaves from stems and seeds from flower heads and spread on trays. If you are unsure about the dry state of mature seeds, place the seeds in an open oven for a few hours. Package whole and grind the seeds as needed. They deteriorate very rapidly when the outer casing is cracked. Properly packaged, the seeds will keep for at least one year without flavor loss. Use sparingly; the flavor is concentrated.

Balm

Balm has a definite lemon odor and flavor. Its leaves are used in many ways—crush a few leaves into poultry dressing or add to ice teas, punches, wines, soups, or desserts. Rubbed on steaks, it adds a pleasant tinge of flavor. It is sometimes used as a tonic and to induce perspiration. The infusion is prepared by covering a small quantity of balm leaves with a pint of water and letting it stand for 15 minutes. If mixture is too strong, add water to dilute.

Lemon balm may be cut four times during the growing season. Leaves should be picked early in the day, after the dew is gone and when the day promises to be hot and dry. Wash carefully as oils may be washed away during the rinsing process. Remove any blemished or bruised leaves. Place on trays in a 100° F. oven with the oven door completely open. Be sure the air is circulating freely. Leaves should take just a few hours to dry. Check every 15 minutes. Turn leaves for faster drying. Rotate trays frequently. Leaves may be crushed before packaging but will retain their oils better if packaged whole. Store in airtight moistureproof containers to keep flavor freshness. Leaves should be checked for moisture or mold after one week of storage. If mold has developed, leaves are not fit to eat or use in recipes.

3> Basil

The ancient Greeks called basil "the herb of kings." It is highly esteemed for its flavorful and pleasingly scented foliage. It is an annual herb that can be grown everywhere. When fresh, the green-leaved basil has a slight licorice taste and when dried a pleasant scent of lemon and anise seed. The dark opal (purple-leaved) basil is less flavorful but can also be used.

Sweet basil is used in eggs, soups, salads, salad dressings, sauces, meats, stews, spaghetti, cheese, and tomato dishes. Wash carefully and pat dry. Remove any blemished or bruised leaves. Place on trays in a

100° F. oven with the oven door completely open. Be sure the air is circulating freely. Leaves should take just a few hours to dry. Check every 15 minutes. Turn leaves for faster drying. Rotate trays frequently. Leaves may be crushed before packaging but will retain their oils better if packaged whole. Package in airtight moistureproof containers to keep flavor freshness. Leaves should be checked for moisture or mold after one week of storage. If mold has developed, leaves are not fit to eat or use in recipes.

Bay Leaf

In ancient times, people were honored with a crown of bay or laurel leaves. Today, the bay leaf is only popular as a seasoning for soups, stews, chicken, fish, meat sauces, marinades, pickles, and in any tomato dish. Its aromatic flavor is strong, so use cautiously.

Wash carefully and pat dry. Remove any blemished or bruised leaves. Place on trays in a 100° F. oven with the oven door completely open. Be sure the air is circulating freely. Leaves should take just a few hours to dry. Check every 15 minutes. Turn leaves for faster drying. Rotate trays frequently. Leaves may be crushed before packaging but will retain their oils better if packaged whole. Package in airtight moisture-proof containers to keep flavor freshness. Leaves should be checked for moisture or mold after one week of storage. If mold has developed, leaves are not fit to eat or use in recipes.

A few dried bay leaves when stored in flour and other dried foods is thought to retard insect infestation.

Black Pepper

Pepper comes from the whole black peppercorn when it has been ground and packaged for table use. Check under *Peppercorns*.

⑤ Borage

Borage is an aromatic herb that resembles cucumber in smell while its delicate flavor resembles oysters in taste. The tender young leaves are sometimes used as a garnish in cool drinks. It can also be served with cauliflower, salads, and stews. The blue flower spikes add beauty as well as flavor when used as a garnish.

To dry leaves: Wash carefully and pat dry. Remove any blemished or bruised leaves. Place on trays in a 100° F. oven with the oven door completely open. Be sure the air is circulating freely. Leaves should take just a few hours to dry. Check every 15 minutes. Turn leaves for faster drying. Rotate trays frequently. Leaves may be crushed before packaging but will retain their oils better if packaged whole. Package in airtight moistureproof containers to keep flavor freshness. Leaves should be checked for moisture or mold after one week of storage. If mold has developed, leaves are not fit to eat or use in recipes.

To dry flower heads: Flower heads are dried whole, and if only the petals are used, remove them from the calyx and follow the instructions given above for drying leaves.

6 Burnet

Burnet is a member of the rose family. Its young leaves are delightful in greens and vegetables. They have a tasty cucumber flavor. Both the leaves and seeds are used to flavor vinegar.

To dry leaves: Wash carefully and pat dry. Remove any blemished or bruised leaves. Place on trays in a 100° F. oven with the oven door completely open. Be sure the air is circulating freely. Leaves should take just a few hours to dry. Check every 15 minutes. Rotate trays frequently. Turn leaves for faster drying. Leaves may be crushed before packaging but will retain their oils better if packaged whole. Package in airtight moistureproof containers to keep flavor freshness. Leaves should be checked for moisture and mold after one week of storage. If mold has developed, leaves are not fit to eat or use in recipes.

To dry seeds: Seed heads should be cut as soon as the seeds are dry. This condition is noticeable in some plants when the seeds or pods change color. This usually occurs just before they scatter. If you are unsure about the dry state of mature seeds, place the seeds in an open oven for a few hours. Package whole and grind the seeds as needed. They deteriorate very rapidly when the outer casing is cracked. Properly packaged, the seeds will keep for at least one year.

7 Caper

Caper is a tiny, greenish, unopened flower bud which can be pickled in salt and vinegar to make a meat sauce. The seeds and fruit have a sharp bitter taste and

120

are used to sharpen flavor in appetizers, pickles, salad dressings, and other condiments. Sometimes pickled nasturtium seeds are sold as capers.

Seed heads should be cut as soon as the seeds are dry. This condition is noticeable in some plants when the seeds or pods change color. This usually occurs just before they scatter. If you are unsure about the dry state of mature seeds, place the seeds in an open oven for a few hours. Properly packaged, the seeds will keep for at least one year.

Capsicum

Capsicum is a genus of plants from the nightshade family, known as peppers. Their berries grow in many sizes, shapes, and colors, ranging from a mild to an extremely hot taste. Included under capsicum are cayenne pepper, chili, and paprika. Check under the individual listings for instructions.

Gloves should be worn during handling of the hot pods and care must be exercised not to touch the body as this might cause burning or scalding that will reactivate if water is inadvertently used on that portion of the body. Dry outdoors. The peppers may be ground after drying.

A tea brewed with cayenne pepper is a good remedy to use on insects and vermin.

Caraway

Caraway is an herb of the carrot family. The young shoots and tender leaves are sometimes used to flavor salads. The plant is biennial and does not set its ovate seeds until the second year of growth. The seeds have an aromatic fragrance and a warm sharp taste. Caraway is used in breads, cheeses, and pastries, and to improve the flavor of some meats. Its spicy oil is used in distilling certain liqueurs such as kümmel. It is sometimes used as a stimulant and a carminative.

Clean seeds. Place on trays in an open oven at a low temperature until dry. Package carefully.

 ## Catnip

Catnip is a hardy, pungent herb of the mint family with mild-tasting leaves. The dried leaves may be crumbled for use in a pleasant savoring tea. Steep for about 10 minutes. Catnip tea is sometimes used for colds and headaches. Cats are especially delighted with the leaves and are attracted to both the fresh and dried catnip.

Wash carefully and pat dry. Remove any blemished or bruised leaves. Place on trays in a 100° F. oven with the oven door completely open. Be sure the air is circulating freely. Leaves should take just a few hours to dry. Check every 15 minutes. Rotate trays frequently. Turn leaves for faster drying. Leaves may be crushed before packaging but will retain their oils better if packaged whole. Package in airtight moisture-proof containers to keep flavor freshness. Leaves should be checked for moisture or mold after one week of storage. If mold has developed, leaves are not fit to eat or use in recipes.

11. Cayenne Pepper

Cayenne pepper is a hot, pungent condiment—a mixture of several kinds of seeds and pods of the capsicum family grown in the Cayenne district of Africa. It is not related to the regular pepper and its color is not as bright as paprika. It is extremely biting and should be used sparingly in chili con carne, tamales, soups, sauces, fish, eggs, cheese, and chicken dishes.

12. Chervil

Chervil is another herb belonging to the carrot family. While its flavor faintly resembles parsley, it is more aromatic and may be substituted for parsley in any recipe. The delicately flavored leaves are used to season appetizers, soups, salads, and sauces. They are especially delectable in cheese, egg, and beef dishes.

Wash leaves carefully and pat dry. Remove any blemished or bruised leaves. Place on trays in a 100° F. oven with the oven door completely open. Be sure the air is circulating freely. Leaves should take just a few hours to dry. Check every 15 minutes. Rotate trays frequently. Turn leaves for faster drying. Leaves may be crushed before packaging but will retain their oils better if packaged whole. Package in airtight moistureproof containers to keep flavor freshness. Leaves should be checked for moisture or mold after one week of storage. If mold has developed, leaves are not fit to eat or use in recipes.

13 Chicory

The leaves of the seedlings are used as greens in salads, while the long cone-shaped roots are boiled like turnips. The chicory has a bitter flavor and while it can be used as a coffee substitute, it is usually used as an added ingredient to coffee.

Wash thoroughly and cut scraped roots into lengthwise slices and boil for several minutes. Place on trays and roast in an open oven at 300° F. until dry and crisp. Grind and store carefully in a cool, dry place.

14 Green Chilies

The long, narrow California (Anaheim) chilies when young and green are used for chili rellenos and other delicious Mexican dishes. The same chili, in its older state, turns red and is then dried for use in making tamales and other Mexican food. The chilies are prepared by washing and simmering gently in a small amount of water until tender. The complete mixture is then sieved to extract the red sauce.

The long chilies in their green or red state may be simply washed and dried whole in a low, open oven. Seeds should be removed. Package carefully when dry. To remove outer skin: Roast, place in paper bag, then refrigerate while warm. Skins should slip off easily when cold.

124

also (14) Chili Powder

Certain types of mature red chilies are ground into chili powder. Its flavor is pungent and fiery and gives zest to Mexican dishes, meats, sauces, tomatoes, corn, eggs, and cheese.

15 Chives

The chive is a member of the onion and leek family. The plant grows about 6 to 8 inches high and its slender, hollow leaves are chopped fine for use as a garnish. Chives are delicious in omelets, salads, soups, and stews. They have a mild onion flavor. This plant is sometimes used for ornamental purposes, as it has tiny delicate lavender flowers.

Wash leaves carefully and pat dry. Remove any blemished or bruised leaves. Place on trays in a 100° F. oven with the oven door completely open. Be sure the air is circulating freely. Leaves should take just a few hours to dry. Check every 15 minutes. Rotate trays frequently. Turn leaves for faster drying. Leaves may be crushed before packaging but will retain their oils better if packaged whole. Package in airtight moistureproof containers to keep flavor freshness. Leaves should be checked for moisture or mold after one week of storage. If mold has developed, leaves are not fit to eat or use in recipes.

16 Clary

Clary is a strongly scented garden herb that has a

125

combined flavor of heavy mint and wild sage. The leaves and flowers are used as seasonings in many foods and drinks. A tea may be made from the fresh or dried flowers. The leaves are also used to replace hops in beer and ale.

Wash leaves carefully and pat dry. Remove any blemished or bruised leaves. Place on trays in a 100° F. oven with the oven door completely open. Be sure the air is circulating freely. Leaves should take just a few hours to dry. Check every 15 minutes. Rotate trays frequently. Turn leaves for faster drying. Leaves may be crushed before packaging but will retain their oils better if packaged whole. Package in airtight moistureproof containers to keep flavor freshness. Leaves should be checked for moisture or mold after one week of storage. If mold has developed, leaves are not fit to eat or use in recipes.

 ## Coriander

Coriander is a small plant herb belonging to the parsley family. The spicy seed somewhat resembles the combined flavors of cumin and curry although its use is different. The dried aromatic seeds are used in breads, candies, cookies, liquors, salad greens, sauces, and seasonings. For a change of pace, add a single seed to a demitasse.

The seed heads, when dried, must be shaken hard to loosen seeds.

18.) Cumin

Cumin is an annual herb of the Old World and a member of the carrot family. Its seeds are aromatic and sometimes bitter. It is used in canapes to stimulate the appetite. Its strong seeds should be used lightly in meat. In its powdered form it is one of the ingredients used in curry and chili con carne, and in flavoring cheese, pickles, sauerkraut, sausages, soups, and stuffed eggs. It can be used liberally in rice dishes.

19.) Curry Powder

Curry is called "the salt of the Orient." It originated in India and is a blend of ten or more spices. Its base is the pulp of the támarind pod and its yellow color is due to the use of large amounts of turmeric. It is usually bought in a powder form although you can grind your own combination. Its seasoning adds an Eastern touch to creamed chicken, tomato soup, veal, lamb, shrimp, chicken, and scrambled eggs. When using curry powder in main dishes, simply serve the dish on rice, with a fruit or iced dessert to complete the meal.

20.) Dill

Dill is a tangy European plant that belongs to the carrot family. Dill seeds resemble caraway seeds in flavor and can be substituted in recipes. Dill seeds are sometimes used to stimulate the appetite.

The young minced leaves are used in salads, as a flavoring in soups and stews, and add fragrance to potatoes, fish, and meats.

Leaves and seeds are both used in seasoning pickles. Dill seeds are especially delicious in borsch, excellent when added to a combination of celery and tomatoes, and zesty in cabbage, cauliflower, and sauerkraut.

Sort seeds. These can be dried in the oven by just the pilot light or at a very low oven temperature with the oven door completely open. Package carefully in airtight containers.

To dry leaves: Wash leaves carefully and pat dry. Remove any blemished or bruised leaves. Place on trays in a 100° F. oven with the oven door completely open. Be sure the air is circulating freely. Leaves should take just a few hours to dry. Check every 15 minutes. Rotate trays frequently. Turn leaves for faster drying. Leaves may be crushed before packaging but will retain their oils better if packaged whole. Package in airtight moistureproof containers to keep flavor freshness. Leaves should be checked for moisture or mold after one week of storage. If mold has developed, leaves are not fit to eat or use in recipes.

Fennel

Sweet fennel is a hardy aromatic herb that belongs to the carrot family. It has the taste and fragrant aroma of anise and both its leaves and seeds are used to flavor candies, cookies, pastries, salads, sweet pickles, and fish. Seeds should be merely picked over and dried.

To dry leaves: Wash leaves carefully and pat dry. Remove any blemished or bruised leaves. Place on

128

trays in a 100° F. oven with the oven door completely open. Be sure the air is circulating freely. Leaves should take just a few hours to dry. Check every 15 minutes. Rotate trays frequently. Turn leaves for faster drying. Leaves may be crushed before packaging but will retain their oils better if packaged whole. Package in airtight moistureproof containers to keep flavor freshness. Leaves should be checked for moisture or mold after the first week of storage. If mold has developed, they are not fit to eat or use in recipes.

 Garlic

Garlic is a hardy perennial bulb that separates into small cloves, protected by a papery skin. Garlic is used sparingly as it has a strong flavor, with a very penetrating smell. Garlic is either prized or disdained and a little goes a long way. It is used to season bread, cheese, meat, pickles, sauces, salads, soups, spaghetti, stews, tomato, and vegetable dishes.

To dry: Peel off skins, slice, place on trays and into a very low open oven until dry and brittle. Bottle or store in covered containers.

To avoid clinging odor, rinse utensils in cold water and then in hot soapy water. Lemon juice may also be used as a deodorizer.

Garlic salt: Cut fresh clove into a few pieces and mix with 1 cup of salt. Stir occasionally and when salt is pungent enough, remove garlic pieces. Store salt in airtight container.

23 Horseradish

Horseradish is a hardy, large-leafed European perennial herb whose grated roots are used to make a pungent food condiment. This peppery root has little food value and is used primarily as an appetizer and to add piquancy to foods. Its strong flavor is sometimes cut with grated beets and sometimes grated turnip is used as an additive. It can be thinned with vinegar, stretched with sour cream, and combined with prepared mustard to use with meats. Combine grated horseradish and cream cheese for a sparkling, biting canape spread. It is commonly used in potato salad and cocktail sauce. Use cautiously. Horseradish will take your breath away and cause your eyes to tear.

24 Hyssop

Hyssop is a member of the mint family whose spicy, pungent leaves and violet-blue flowers are used to flavor soups, salads, beverages, candies, and in making absinthe as well. It is also used as a medicinal herb for coughs and colds.

Wash leaves carefully and pat dry. Remove any blemished or bruised leaves. Place on trays in a 100° F. oven with the oven door completely open. Be sure the air is circulating freely. Leaves should take just a few hours to dry. Check every 15 minutes. Rotate trays frequently. Turn leaves for faster drying. Leaves may be crushed before packaging but will retain their oils better if packaged whole. Package in airtight moistureproof containers to keep flavor freshness. Leaves should be checked for moisture or mold after the first week of storage. If mold has developed, they are not fit to eat or use in recipes.

Mace

nutmeg

Mace comes from a tropical fruit of an evergreen grown in the East and West Indies. It is the lacy aril that protects the hard inner shell of the nutmeg kernel. Its fragrance is similar to nutmeg but dissimilar in use and flavor. It is used in baking and to enhance the color and flavor of cakes. For a touch of elegance—add a teaspoon of mace to a pint of whipping cream.

Marjoram

Sweet marjoram is a highly aromatic perennial herb of the mint family. It is sometimes confused with wild marjoram, which is oregano. This highly spiced herb is grown for its fragrant foliage which is used in flavoring meat and dressing. It should always be used with lamb and mutton. It blends well with fish, cheese dishes, soups, stews, and sauces, and adds a touch of zest to vegetables. It can be used fresh or dried but should be used sparingly as it has a strong flavor. It harmonizes and blends extremely well in whatever food it is in.

Wash leaves carefully and pat dry. Remove any blemished or bruised leaves. Place on trays in a 100° F. oven with the oven door completely open. Be sure the air is circulating freely. Leaves should take just a few hours to dry. Check every 15 minutes. Rotate trays frequently. Turn leaves for faster drying,

Leaves may be crushed before packaging but will retain their oils better if packaged whole. Package in airtight moistureproof containers to keep flavor freshness. Leaves should be checked for moisture or mold after the first week of storage. If mold has developed, they are not fit to eat or use in recipes.

For flowers and stems follow above outline.

Mint

Mint is an aromatic herb belonging to the genus *Mentha*. It has several varieties including spearmint, peppermint, and horsemint. Mint is usually fragrant with a very agreeable flavor. It is used in jellies, candies, and fruit salads, and its sprigs are sometimes used as a garnish. Minced, it enlivens rice, peas, and carrots. It can also be used as a fly repellent. Hang the sprigs over doorways or place into areas where fleas and flies are troublesome.

Nutmeg (mace)

Nutmeg is the fruit of a species of a tropical evergreen tree of the East and West Indies whose single fruit splits to produce two spices. The lacy aril is commonly known as mace and it protects the hard inner shell of the nutmeg kernel. Nutmeg is usually sold in powder form, although whole nutmegs can be bought. The freshly ground powder has more flavor. Its spice adds a touch of elegance to eggnogs, custards, and rice puddings. It heightens flavor when used on meats, fish, poultry, and cauliflower.

29. Oregano

Oregano is the Spanish name for wild marjoram and sometimes is confused with sweet marjoram. Both are used in many of the same foods but they are not exchangeable. Oregano is more potent than sweet marjoram. This zesty herb is used to flavor tomato and vegetable juices, eggs, spaghetti sauces, pizzas, meats, and many Italian dishes. It is an important ingredient in many chili con carne recipes. Oregano may be substituted for thyme in any recipe.

30. Paprika

Paprika is obtained by grinding milder pods of the capsicum family. The most excellent grade comes from Hungary and, the brighter the color, the better the quality. A sprinkling of paprika will highlight bland foods and give them a festive touch. For a subtle but fine dressing, coat greens lightly with olive oil and sprinkle with paprika to taste—very different but very tasty. Paprika also makes a good browning agent for roasts and other meats. Try it on seafood for a nice touch and flavor.

31. Parsley

This is an aromatic herb used as a garnish and as a seasoning. Fresh or dry, parsley adds distinction to all cooked or prepared foods. The curly-leaf parsley is usually used as a garnish but it does add a delicately mild flavor if used in food. The Italian parsley is grown primarily for its flavor. It has a dark green color, broader leaves, and is more piquant in taste. There is still another form which is grown primarily for its huge, white, parsnip-type root. This is used only for flavoring.

Parsley is very popular in canapes as it is said to stimulate the appetite. Parsley also complements canned salmon and mackerel patties. Add a tablespoon of dried parsley to the sautéed onions before adding to fish. In the olden days, it was thought that a tea steeped from parsley leaves would help rheumatism and stimulate the kidneys. This tea was served cold.

To dry: Leaves should be fresh and green. Wilted or yellow parsley is almost worthless. Sometimes wilted parsley may be revived if placed into cold water. Wash leaves carefully and pat dry. Remove any blemished or bruised leaves. Place on trays in a 100° F. oven with the oven door completely open. Be sure the air is circulating freely. Leaves should take just a few hours to dry. Check every 15 minutes. Rotate trays frequently. Turn leaves for faster drying. Leaves may be crushed before packaging but will retain their oils better if packaged whole. Package in airtight moisture-proof containers to keep flavor freshness. Leaves should be checked for moisture or mold after the first week of storage. If mold has developed, they are not fit to eat or use in recipes.

 Peppercorns

Botanically the peppercorn is the spice of a tropical plant bearing spike-shaped clusters of fruit which hold the round peppercorns.

Centuries ago it was used mainiy as a food preservative and for this reason became a chief source of trade. Pepper was worth its weight in gold and fortunes were won and wars were fought between nations on land and with piracy on the high seas over pepper. Peppercorns were held with such high esteem that porters unloading the precious commodity wore uniforms without pockets to prevent thievery. Very severe penalties were administered to those caught stealing pepper. With faster transportation, pepper and many other spices lost their merchandising appeal.

Clusters are picked just before they turn red. The peppercorns are considered dried when they have turned brown/black and are hard and wrinkled. The berries are stripped off the spikes and packaged for home use. They are used in canning, in soups, stews, and other foods. Use freshly milled peppercorns to add zest to prepared dishes.

33 Poppy Seeds

Tiny black poppy seeds have a distinctive nutty flavor, with just a hint of fragrance. They are used as flavorings and to stimulate the appetite and for this reason they are blended in or sprinkled on canapes. They are excellent on breads, rolls, cookies, and are outstanding in poppyseed cakes. The fillings make rich but delectable pastries. Sprinkle them on noodles, use them in pickles and preserves and add them to vegetable sauces. The oil derived from the seeds is used in salads and for cooking purposes. They do not come from the opium poppy.

Rosemary

Rosemary is a very fragrant evergreen shrub belonging to the mint family. Its leaves have a delightful piney fragrance but are very potent and should be used sparingly. Rosemary can be purchased fresh, dried, or in ground form. Dried rosemary should be dipped into hot water then cold water to liven up its flavor. It can be combined with chives and thyme to complement salads. Rosemary is used to season soups, sauces, fish, lamb, beef, and pork. Try it as an herb tea or in herb soups. Rosemary wine is excellent when added in small quantities to gravies or fish. It has a totally different but subtle flavor. Add a good heaping teaspoon of rosemary to a cup of red or white wine. Keep in a cool place for one week (not the refrigerator). Strain and store in a cool place.

Wash leaves carefully and pat dry. Remove any blemished or bruised leaves. Place on trays in a 100° F. oven with the oven door completely open. Be sure the air is circulating freely. Leaves should take just a few hours to dry. Check every 15 minutes. Rotate trays frequently. Turn leaves for faster drying. Leaves may be crushed before packaging but will retain their oils better if packaged whole. Package in airtight moistureproof containers to keep flavor freshness. Leaves should be checked for moisture or mold after the first week of storage. If mold has developed, they are not fit to eat or use in recipes.

Saffron

Saffron is an extremely sweet smelling annual that is

136

used primarily to color candies, vegetables, and season foods. Saffron is made from the dried stigmas of the autumn crocus and the blossoms must be picked by hand. It is probably the most expensive spice on earth. It is utilized as a condiment in pickles, sauces, meat, fish, and poultry.

36 Sage

Sage is a hardy, shrubby aromatic herb of the mint family whose grayish green leaves are dried and used as a flavoring or seasoning. Sage is generally used with sausages and pork. Botanically, sage is called salvio, which means to save. This is probably because many healing properties have been ascribed to it throughout the ages. Its strong flavor seasons soups, cheeses, stews, sausages, and dressings. It should be stored in airtight containers to keep its pungency. Sage is the second strongest herb, with parsley ranking first. Sage can be replaced by savory in any recipe.

Wash leaves carefully and pat dry. Remove any blemished or bruised leaves. Place on trays in a 100° F. oven with the oven door completely open. Be sure the air is circulating freely. Leaves should take just a few hours to dry. Check every 15 minutes. Rotate trays frequently. Turn leaves for faster drying. Leaves may be crushed before packaging but will retain their oils better if packaged whole. Package in airtight moistureproof containers to keep flavor freshness. Leaves should be checked for moisture or mold after one week of storage. If mold has developed, they are not fit to eat or use in recipes.

Savory
(Summer Savory)

Savory is a hardy annual and a member of the mint family. It is sometimes called the "bean herb" as it is an excellent seasoning when used in bean dishes. It can be used alone or combined with other herbs. It adds a very delicate flavor to soups, salads, dressings, stuffings, poultry, and meats. It can also be used as a substitute for sage in recipes.

Wash leaves carefully and pat dry. Remove any blemished or bruised leaves. Place on trays in a 100° F. oven with the oven door completely open. Be sure air is circulating freely. Leaves should take just a few hours to dry. Check every 15 minutes. Rotate trays frequently. Turn leaves for faster drying. Leaves may be crushed before packaging but will retain their oils better if packaged whole. Package in airtight moistureproof containers to keep flavor freshness. Leaves should be checked for moisture or mold after a week of storage. If mold has developed, they are not fit to eat or use in recipes.

Sesame Seeds

Sesame seeds are cultivated for their oil which is used in cooking and in salad dressings. These little golden nuggets also add a festive touch to bread, rolls, biscuits, cakes, cookies, and candies and a touch of elegance to noodles, potatoes, meat, and poultry. Their delightful flavor is especially valued as a condiment. Toast in oven at 350° F. for 12-15 minutes, turning frequently, until golden brown.

34 Shallots

The shallot is a small, mild, pear-shaped member of the onion family but more aromatic and used primarily for the flavoring which combines some of the best traits of onions, garlic, and scallions. Its light outer skin is reddish or gray and the inner portion is green at the top and purple at the bottom. It is sometimes used as a substitute for onion and in some recipes onions and shallots are used together. Shallots are usually used fresh for flavoring in various sauces.

To dry: Follow recipe for *Garlic.*

35 Spearmint

A hardy perennial aromatic herb that is used for flavoring. It is sometimes referred to as "lamb mint" because its leaves are often used in making a sauce for lamb. Its oil is used as a flavoring extract in medicines, gums, and candies. Spearmint leaves are the main ingredient in making mint juleps.

Wash leaves carefully and pat dry. Remove any blemished or bruised leaves. Place on trays in a 100° F. oven with the oven door completely open. Be sure air is circulating freely. Leaves should take just a few hours to dry. Check every 15 minutes. Rotate trays frequently. Turn leaves for faster drying. Leaves may be crushed before packaging but will retain their oils better if packaged whole. Package in airtight containers to keep flavor freshness. Leaves should be checked for moisture or mold after a week of storage. If mold has developed, they are not fit to eat or use in recipes.

Sunflower Seeds

Sunflowers have characteristic daisy heads with yellow petals and inner colored circles that range from brown to almost black. They are harvested when most of the golden petals have fallen. To release the seeds, rub your hand back and forth across the flower heads. Place them on trays into a very low open oven to dry. Package carefully.

Soak dried seeds for 1 hour in a mixture of 1 cup pickling salt to 2 quarts of water. Drain. Place on cookie sheets and roast at 150° F. for about 2 hours or until dry. Store carefully.

Seeds blend well when added to other nut treats for a delicious snack, or use in cereals and candies. Remember, birds and chickens like them too!

Tarragon

Tarragon is a European herb of the aster family closely related to the wormwood. Its bitter aromatic leaves are used for seasonings and flavorings. Add it to vinegar, fish, gravy, meats, pickles, stuffing, and vegetables. It is a must in lobster thermidor and Bearnaise sauce. Just use it sparingly!

To dry: Follow outline for leaves.

Thyme

Thyme is another member of the mint family that is valued for its seasoning. It is sometimes used as an ornament in rock gardens or as an herb border. Some

140

species are cultivated for the oil in their leaves and stems. The leaves are pungent and aromatic and it is a favorite among herbs. It is an essential ingredient in bouquet garni. Use it to season meats, fish, poultry, soups, stews, stuffings, and to heighten the flavor of some vegetables.

Wash leaves carefully and pat dry. Remove any blemished or bruised leaves. Place on trays in a 100° F. oven with the oven door completely open. Be sure air is circulating freely. Leaves should take just a few hours to dry. Check every 15 minutes. Rotate trays frequently. Turn leaves for faster drying. Leaves may be crushed before packaging but will retain their oils better if packaged whole. Package in airtight moistureproof containers to keep flavor freshness. Leaves should be checked for moisture or mold after one week of storage. If mold has developed, they are not fit to eat or use in recipes.

Turmeric

Turmeric is derived from a tropical plant related to the ginger family. Its aromatic rootstock is powdered and used as a dye to tint butter, cheese, pastries, and other foods. It is also used to flavor mustard, pickles, and condiments.

White Pepper

(See Peppercorns)

White pepper is the inner section of the peppercorn. Its outer portion has been removed by soaking or buffing. It is less pungent and is used to season and blend delicately with foods.

chpt
10.

Nuts

goes from 143 to 147. 5 pgs

Nuts are an excellent source of protein and contain rich body-building nutritives. They are eaten primarily before a meal to stimulate the appetite. Sometimes they are sprinkled on vegetables for added protein or to highlight color and to emphasize flavor. Nuts should be avoided after a rich meal as the added nutrients may cause indigestion.

Two ounces of nuts, when served along with other foods, will help balance a diet. Nut meats may be combined with other seeds and kernels for a delightful snack. Some nuts, such as hickory nuts, black walnuts, and English walnuts should not be added to salads until just before serving. The nut meats will darken and permeate the whole mixture, resulting in a very unappetizing product.

Mature nuts should be harvested as soon as they have fallen. Mold sets in very quickly, especially in damp weather. This is particularly true of the English walnuts and chestnuts. A constant, vigilant harvesting will result in less spoilage and greater yields.

A simple way of sorting out diseased, rotten, or wormy nuts is to merely place them in water. The defective nuts will rise to the surface where they can be easily removed. Soft-shelled nuts may be easily peeled by cutting across the base with a knife, then cutting off the rind with the knife end. Shells should slip off easily.

Black walnuts need prompt attention as soon as they are harvested. They should be hulled within a week in order to keep their mild flavor and light color. The shells exude a bitter flavor which will permeate through to the kernel. This will discolor and detract from their flavor. This acrid fluid will also stain badly, so protect your hands and skin.

Nuts need an exposure to heat and air to reach their prime condition. Circulating air is of utmost importance. Carelessness in either will result in an inferior product that is tasteless, tough, and wrinkled.

Blanching: Shelled nuts are easy to skin. Merely drop them into boiling water for 5 minutes. Rinse under running water and gently slip off the skins. Butter, oil, or seasonings cause rancid nuts and should be added just before serving.

Dried nuts should be packaged immediately. If not exposed to air, light, or moisture they should keep for about two years.

Almonds

Drying time should start within 24 hours of harvest and the excess held in a shady, airy place. Ripeness is indicated by splitting hulls. The shells and kernels are beginning to dry out.

Remove hulls and place almonds on trays and into the oven at 115° F. The oven door on a gas range must be open 8 inches, on an electric range half an inch for air circulation. Dry for 12-24 hours or until almonds are crisp.

When packaging, remove as much air as possible, since it is the air that causes the deterioration. Almonds will keep for several years if dried and packaged properly. Use airtight containers. Almonds may be blanched to remove skins. See the instructions given in the section *Nuts*.

2> Chestnuts

Chestnuts are small, sweet, and delicious boiled, roasted, or pureed. They are excellent when served with Brussels sprouts. Chestnuts do not have to be dried artificially as they will eventually dry naturally in their own shells.

For the more enterprising and eager, make a gash in the shell with a sharp knife and place chestnuts into boiling water for about 10-15 minutes. Drain and peel off shells and skins immediately.

Place on trays into a 140° F. oven and dry until hard and shriveled. The oven door on a gas stove must be open 8 inches, on an electric stove half an inch. Package carefully.

To reconstitute: Simply soak them in water.

3> Coconut

The remarkable coconut tree is a warehouse of treasures—every portion can be used in some way. Its fruit provides milk, food, astringents, antiseptics, and medicines. The leaves are used for rooftops and for lighting. Mattresses are made from the fibers of the coconut husks, the shells are used for eating and drinking, while the wood is used in making furniture and building houses. The roots are ground for tea or coffee.

Coconuts are easily dried and their tasty shredded meat adds a festive touch to any dish. Open the hard shell, remove the meat, and trim off the brown skin. It can be left in pieces or grated and dried in a very low oven with the door open. Or open the hard shells and place in the oven to dry until the meat separates from the shell. Remove and continue to dry until meat is hard, crisp, and creamy white. It can then be grated or packaged whole. *Tap the "eyes" on the coconut first with a hammer & drain the liquid ...*

 ## Filberts

Filberts are nuts from a bushy shrub or tree of the Oriental hazel. They may be harvested after they have fallen from the tree. While the quality of the nuts are not affected by this, they should be processed within 24 hours to reserve their peak quality. Filberts, like other nuts, need low heat and good air circulation. They may be dried in the shell and placed into a cool storage area for short intervals or into the freezer for long-term storage. Filberts are used in making delicious confections.

Peanuts

A tender annual of the legume family, peanuts are sometimes called monkey nuts, goobers, pindar earth nuts, or ground nuts. They have many other uses other than making peanut butter and roasted nuts that we identify with. Peanut flour is made from the ground nuts and a type of peanut meal is sometimes used by diabetics as it has little or no carbohydrates. Peanut oil is a very important product, and outside of its usefulness in the kitchen, it is sometimes used in making soap and preparing oleo. The plant is fodder for the animals while the ash of the shell is used for fertilizers. Spread nuts on trays and place in the oven at low heat until dry and crunchy. The oven door on a gas range must be open 8 inches, on an electric range half an inch.

⑥ Walnuts

Walnuts were used in trade in early history, and because English trading ships transported them all over the world, they were misnamed and called English walnuts. The nuts reach maturity in the fall, or as soon as the husk will cut away from the shells. Walnuts should be picked several times during the harvesting season to prevent loss through decay or mold. These creamy wrinkled nuts lose their quality quickly after they have dropped and should be dried within 24 hours. Whole nuts are dry when the center divider will break cleanly. They should be stored in a cool, dark place. Shelled nuts take less time to dry and should be stored in the freezer for longer keeping qualities. Low heat and good air circulation is most important.

Black walnuts are extremely delicious but it takes 150 years for the trees to reach maturity. It is wonderful and fortunate that the Carpathian variety takes only 6-8 years.

11.

A Little Bit of This and That

The following recipes and instructions on the preparation of miscellaneous foods will include candies, granola, cookies, eggs, pectin, and vinegar. I hope that these will prove rewarding in terms of work and purse. All are simple in outline and preparation—outstanding in terms of nutrition, taste, and quality.

Candy

Candy is a highly nutritious and excellent convenience food when camping, hiking, etc. These recipes are very simple to prepare and are delicious for a healthful, quick pick-me-up that is sometimes necessary when either the time or place is inconvenient for the regular meal time schedule.

1. HONEY BARS. —
¼ cup honey
¾ cup instant nonfat dry milk
½ cup chopped walnuts
½ cup chopped dates

148

½ cup chopped raisins
½ cup shredded coconut
¼ teaspoon orange extract
¼ cup grated orange peel

Stir honey and milk together until well blended. Add other ingredients and mix well. Form into two oblong rolls. Cover with plastic wrap and chill in refrigerator.

② FRUIT CANDY. —
2 tablespoons powdered pectin
¼ cup corn syrup
1¼ cups of any fruit juice
2 tablespoons cornstarch
2 cups sugar
Optional: ¼ cup chopped seeds or nuts

Mix together pectin, corn syrup, and 1 cup juice, reserving ¼ cup juice. Bring slowly to a boil, stirring constantly for 2 minutes. Add cornstarch to remaining juice, blend, then add to simmering mixture and stir for 2 more minutes. Add sugar and cook for 12 minutes, adding any seeds or nuts during the last few minutes of cooking time. Turn out into a 7 by 12 inch baking dish. Cool, then refrigerate and cut into squares when cold. Sprinkle with powdered sugar if desired.

③ FRUIT PASTE. —
2 cups any fruit puree
2 cups sugar
Optional: Finely chopped nuts, sunflower seeds, orange, or lemon peel

Combine puree and sugar and cook over low heat, stirring mixture frequently until thick. Cook over boil-

C
A
N
D
Y

ing water for 30 minutes. Nuts, seeds, or peels may be added at this time. Pour onto a plastic-lined cookie sheet and place in an open oven at a very low temperature until firm. Cut into pieces or roll into sugar, coconut, or nut meats. Pack into boxes.

④ Granola

This recipe has been included as it is an excellent way of utilizing the fruits and nuts prepared earlier in the season and contains vitamins and nutrients so necessary in this age of bleached foods and food additives. Granola is extremely good for children who need nourishing treats that are filled with vitamins and excellent in taste and quality.

¼ cup oil
¼ cup honey
3 cups quick oatmeal
½ cup chopped walnuts or other nuts of your choice
½ cup chopped dates or other dried fruits or seeds
½ cup chopped raisins

Heat oil and honey together over low heat, stirring carefully until warm and blended. Pour over oatmeal and mix until oatmeal is well coated. Place in a 7 by 12 inch baking pan and into a 325° F. oven for 25-30 minutes. Stir ingredients twice during the baking time. Remove from oven and mix in other ingredients.

This cereal may be prepared for mealtime use or served as a delicious snack food.

Cookies

This very simple recipe is brimful of health boosters. It is nutritious in terms of the vitamins and minerals it contains and delicious in terms of taste—an aid to health through healthful snacking. This is another excellent convenience food to take along on outings away from home.

⑤ HEALTH BARS. —

¼ cup oil
¼ cup honey
3 cups quick oatmeal
½ cup chopped walnuts or other nuts of your choice
½ cup chopped dates or other dried fruits or seeds
½ cup chopped raisins

Heat oil and honey together over very low heat, stirring carefully until warm and blended. Pour over oatmeal and·mix and stir until well coated. Add other ingredients and form two oblong rolls on a cookie sheet. Bake at 325° F. for 25-30 minutes until golden brown. Do not attempt to remove bars until they are cold as they will crumble. Use a serrated knife to cut into slices.

Eggs

Eggs should be purchased when the prices are low and the market abundant. The instructions are easy to follow, making them convenient for later use when prices are high or eggs are scarce. Dried eggs are very handy —they add more nutrition to foods when used as a seasoner, leavener, or simply as a decorative item, sprinkled on foods. Included are instructions on drying the whole egg—a first in home drying!

Only fresh eggs should be used. The outer shell should appear dull and free from cracks and any veining. Fresh eggs sink when placed into cold water. Chilled eggs separate more easily, while whites beat up better at room temperature. Eggs are broken into separate bowls before adding to the main bowl.

(1) DRIED EGG WHITES. —

Make a stiff meringue with 12 egg whites and 1 teaspoon cream of tartar. Spread meringue *evenly* on cookie sheet that has been lined with aluminum foil and place in a 225° F. oven for ½ hour then open oven to reduce temperature to 140° F. for another hour or until crisp and dry. Moisture content should be 3 percent when dry. Powder with a rolling pin.

(2) DRIED EGG YOLKS. —

Whip 12 yolks until thick like heavy cream and very light in color. Spread evenly on a cookie sheet that has been lined with aluminum foil and place in 150° F. oven for 5 minutes. Lower temperature or open door to reduce temperature and continue until dry (about 1½ hours or 3 percent moisture content). If mixture appears to be dry but is still moist, return to the oven to finish. Powder with a rolling pin. (If mixture has not been whipped enough, it will form a set finish. Place in blender to powder after it is dried.)

(3) DRIED EGG. —

Make a stiff meringue of 6 egg whites and ½ teaspoon cream of tartar. Beat yolks in a separate bowl until thick like cream and light in color. Carefully spoon some of the yolk mixture into the whites, folding carefully. Add balance of yolks gradually until all is blended. *Do not stir!* Spread evenly on a cookie sheet that has been lined with aluminum foil. Place into preheated oven at 350° F. for 5 minutes. *Do not allow mixture to brown.* Lower oven temperature to the lowest possible point, leave the oven door completely open, and allow mixture to remain until completely dried. Dry for 1½ hours or 3 percent moisture content when dry. Powder with a rolling pin.

PACKAGING.—

It is absolutely *essential* that the dried egg whites, dried yolks, or whole dried eggs be packed in *airtight, moistureproof* bags or containers. If in doubt, destroy product.

RECONSTITUTION. —

To 1 firmly packed tablespoon egg white and 1 firmly packed tablespoon egg yolk gradually add 3 tablespoons *warm* water. This will equal 1 egg. Remember to introduce a small amount of water and blend into a paste before adding the rest of the water.

To 2 firmly packed tablespoons dried whole egg gradually add 3 tablespoons warm water. This will equal 1 egg. Introduce liquid gradually and blend into a paste before adding liquid balance.

If added to recipes, the amount of water required to reconstitute the dried eggs should be included in the recipe.

Pectin

Pectin is a carbohydrate substance found in pulpy

fruits which cause fruit or juice to jell when boiled. It is usually found in greater proportions in underripe fruit than in those that are fully ripe. An overabundance of this substance can result in green-apple colic.

Pectin may also be processed in a water bath for later use. Once opened, it must be used as it will not keep.

The following recipe is an excellent way of using sound but small, knarled, or misshapen apples.

APPLE PECTIN. —

7 large tart apples or 2 pounds of apples
4 cups water
3 tablespoons lemon juice or the juice of 1 lemon

Wash apples. Stem, remove any imperfections, and cut into small pieces. Combine with juice and water; simmer for 40 minutes. Press through a jelly bag, then boil rapidly for 15 minutes. Pour into sterilized jars and seal. Process in water bath for 20 minutes.

Use 1 cup apple pectin with 1 cup fruit juice and 3/4 cup sugar. Cook until fruit jells and pour into sterilized containers.

LEMON OR ORANGE PECTIN. —

The white portion of the lemon or orange can be easily converted into pectin. Remove the outer yellow or orange peel and grind the white portion. To each ½ cup of ground product (¼ pound), add 2 tablespoons lemon juice and 2½ cups cold water. Let stand 2 hours. Bring to a boil and cook until amount is reduced by half. Strain and reserve liquid. Repeat this cooking process two more times, omitting the 2 hours of soaking time. Combine the 3 liquids and bring to a boil. Pectin may be used immediately or may be poured into sterilized jars and processed in a water bath for 30 minutes.

Use 2 cups fruit juice, 2 cups sugar, and 2/3 cup lemon or 3/4 cup orange pectin. Boil rapidly until the fruit juice jells, then pour into sterilized containers.

Vinegar

Variety may be made from almost any tart fruit and it is a wonderful way of. using blemished (not spoiled) fruit that cannot be eaten or canned. The most popular is the apple but other fruits may also be used. Cut fruit into pieces, adding the skins, seeds, and cores. Place into crock and mash until fruit is soft and mushy. Cover and keep at a moderate temperature. Age at least six months, stirring occasionally and tasting for vinegar. When it has reached the desired strength, strain and place the liquid into clean jars. Store in a cool, dark place. Do not tighten the lids on the jars.

Vinegar may be pasteurized by submerging sealed bottles in a warm water bath. Heat to 140-160° F. and maintain this temperature for 5 minutes for pints and 10 minutes for quarts.

— The End —

156

goes from pg 156 to 158...

Index

INSECTS

S - (cont.)